# TOPICS IN

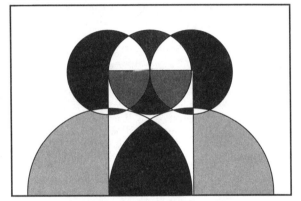

# INTERSECTION
# GRAPH THEORY

# SIAM Monographs on
# Discrete Mathematics and Applications

The series includes advanced monographs reporting on the most recent theoretical, computational, or applied developments in the field, introductory volumes aimed at mathematicians and other mathematically motivated readers interested in understanding certain areas of pure or applied combinatorics, and graduate textbooks. The volumes are devoted to various areas of discrete mathematics and its applications.

Mathematicians, computer scientists, operations researchers, computationally oriented natural and social scientists, engineers, medical researchers and other practitioners will find the volumes of interest.

## Series Volumes

McKee, T. A. and McMorris, F. R., *Topics in Intersection Graph Theory*
Grilli di Cortona, P., Manzi, C., Pennisi, A., Ricca, F., and Simeone, B., *Evaluation and Optimization of Electoral Systems*

# TOPICS IN

# INTERSECTION GRAPH THEORY

## Terry A. McKee
Wright State University
Dayton, Ohio

## F. R. McMorris
University of Louisville
Louisville, Kentucky

Society for Industrial and Applied Mathematics     Philadelphia

**Library of Congress Cataloging-in-Publication Data**

McKee, Terry A.
    Topics in intersection graph theory / Terry A. McKee, F. R. McMorris.
        p. cm. -- (SIAM monographs on discrete mathematics and applications)
    Includes bibliographical references and index.
    ISBN 0-89871-430-3
    1. Intersection graph theory. I. McMorris, F. R. II. Title. III. Series.
    QA166.185.M34   19996
    511'.5--dc21                                          98-31901

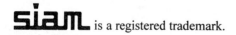 is a registered trademark.

# Contents

# Preface

Intersection graphs provide *theory* to underlie much of graph theory. They epitomize graph-theoretic structure and have their own distinctive concepts and emphasis. They subsume concepts as standard as line graphs and as nonstandard as tolerance graphs. They have real applications to topics like biology, computing, matrix analysis, and statistics (with many of these applications not well known).

While there are other books covering various topics of intersection graph theory, these books have focus and intent that are different from ours. Even those that are out of date are still valuable sources that we urge our readers to consult further. [Golumbic, 1980], with its partial updating in [Golumbic, 1984], remains a standard, excellent source, organized around perfect graphs. There is much related content in [Roberts, 1976, 1978b], both of which emphasize intersection graphs and applications. Among others, [Berge, 1989] develops many of the general concepts in terms of hypergraphs, [Fishburn, 1985] and [Trotter, 1992] stress an order-theoretic viewpoint, [Kloks, 1994] emphasizes treewidth, and [Prisner, 1995] focuses on graph operators. [Mahadev & Peled, 1995] is devoted to threshold graphs. [Brandstädt, 1993] and [Brandstädt, Le, & Spinrad, to appear] discuss many of the relevant graph classes. [Zykov, 1987] includes valuable references to the Russian literature up to that date.

We have tried to write a concise book, packed with content. The first four chapters focus on what we feel are the most developed topics of intersection graph theory, emphasizing chordal, interval, and competition graphs and their underlying common theory; Chapter 5 discusses the allied topic of threshold graphs. Chapter 6 extends the common theory to $p$-intersection, multigraphs, and tolerance. Chapter 7 adopts a different spirit, serving as a guide to an active, scattered literature; we hope it communicates the flavor of various topics of intersection graph theory by offering tastes of enough different topics to lure interested readers into pursuing the citations and learning more. We have pointed in a multitude of directions, while resisting trying to point in all directions.

We have made the book self-contained modulo the basics present in any introductory graph theory text, whether one like [Chartrand & Lesniak, 1996] with virtually no overlap with our topics, or one like [West, 1996] that introduces several of the same topics. We hope it can serve as a platform from which one can launch more detailed investigations of the broad array of topics that involve intersection graphs. The more than one hundred simple exercises scattered throughout the first six chapters are meant to be done as they occur, to reinforce and extend the discussion.

In spite of its size, the Bibliography does not pretend to be complete. Many relevant papers are not included—even some of our own—partly by design and partly reflecting our ignorance and prejudices. We hope that even connoisseurs will find a few surprises, though. We have made a special effort to include early papers and recent papers with good bibliographies, but we have typically included very few papers that emphasize solving particular problems (e.g., coloring, domination, identifying maxcliques, and a host of others) or that emphasize details of algorithms and complexity. Papers marked as "to appear" had not been published when this book was completed and should be looked for using the American Mathematical Society's MathSciNet. We also intend limited updating (including, inevitably, corrections) on a web site locatable though the authors' home institutions.

The following are among the possible uses of this book: (i) as a source book for mathematical scientists and others who are not familiar with this material; (ii) as a guide for a research seminar, utilizing the references to explore additional topics in depth; (iii) as a 5–6 week "unit" in an advanced undergraduate/graduate level course in graph theory.

We acknowledge the valuable input of anonymous reviewers and the encouragement and interest of many colleagues, Peter Hammer in particular. We thank Jenő Lehel in particular for comments on certain portions of the manuscript, while of course we retain all responsibility for lapses and shortcomings.

$$(Mc)^2$$

# Chapter 1

# Intersection Graphs

The goal of this chapter is to present basic definitions and results for intersection graphs of arbitrary families of sets. This machinery will then be used as the basis for the more specialized topics in the following chapters. Much of the viewpoint of this chapter reflects [Roberts, 1985].

## 1.1  Basic Concepts

We follow the standard terminology and notation that is common to most graph theory texts, such as [Chartrand & Lesniak, 1996] or [West, 1996]. For instance, $V(G)$ and $E(G)$ refer respectively to the sets of vertices and edges of a graph $G$ of order $|V(G)|$ and, for $u, v \in V(G)$, $uv$ refers to the edge joining $u$ and $v$. Uncommonly, we allow the *null subgraph* of $G$, meaning the graph $K_0$ having $V(K_0) = \emptyset = E(K_0)$. In particular, the null subgraph is a complete subgraph of every graph (section 4.2 will show one reason why this is desirable).

By a *family* $\{S_1, \ldots, S_n\}$ of sets or graphs we mean a *multiset*, which allows the possibility that $S_i = S_j$ even though $i \neq j$. Unless we specifically say otherwise, all graphs and digraphs will be finite and graphs will have neither loops nor multiple edges.

We define a *maxclique* of a graph to be any complete subgraph that is not properly contained in another complete subgraph. (Warning: some authors use "clique" for what we call "maxclique," while for many others a clique can be *any* complete graph.) For instance, the graph shown on the left in Figure 1.1 has two maxcliques, of orders two and three.

Let $\mathcal{F} = \{S_1, \ldots, S_n\}$ be any family of sets. The *intersection graph of* $\mathcal{F}$, denoted $\Omega(\mathcal{F})$, is the graph having $\mathcal{F}$ as vertex set with $S_i$ adjacent to

1

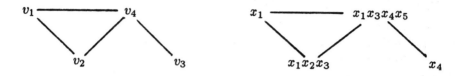

Figure 1.1: *An intersection graph $G$, both "plain" and set labeled.*

$S_j$ if and only if $i \neq j$ and $S_i \cap S_j \neq \emptyset$. A graph $G$ is an *intersection graph* if there exists a family $\mathcal{F}$ such that $G \cong \Omega(\mathcal{F})$, where we typically display this isomorphism by writing $V(G) = \{v_1, \ldots, v_n\}$ with each $v_i$ corresponding to $S_i$; thus $v_i v_j \in E(G)$ if and only if $S_i \cap S_j \neq \emptyset$. When $G \cong \Omega(\mathcal{F})$, $\mathcal{F}$ is then called a *set representation* of $G$.

**Example 1.1** Suppose $\mathcal{F} = \{S_1, S_2, S_3, S_4\}$ where $S_1 = \{x_1\}$, $S_2 = \{x_1, x_2, x_3\}$, $S_3 = \{x_4\}$, and $S_4 = \{x_1, x_3, x_4, x_5\}$. Then $G \cong \Omega(\mathcal{F})$ is shown in Figure 1.1. It is sometimes useful to label the vertices of an intersection graph $G$ with the actual sets of $\mathcal{F}$ (abbreviating $\{x_1, x_2, x_3\}$ as $x_1 x_2 x_3$, etc.), producing the graph on the right in Figure 1.1, which we call a *set-labeled intersection graph.*

Suppose $G \cong \Omega(\mathcal{F})$ where $\mathcal{F} = \{S_1, \ldots, S_n\}$ and each $v_i \in V(G)$ corresponds to $S_i \in \mathcal{F}$ under the isomorphism. For each $x \in \cup_{i=1}^n S_i$, set $G_x = \{v_i : x \in S_i\}$. It is easy to see that each $G_x$ induces a complete graph of $G$ of order $|\{i : x \in S_i\}| \geq 1$.

**Example 1.1 (continued)** For the given family $\mathcal{F}$ and $G \cong \Omega(\mathcal{F})$, $G_{x_1} = \{v_1, v_2, v_4\}$ (these being the vertices corresponding to the three $S_i$'s that contain $x_1$); similarly $|G_{x_2}| = 1$, $|G_{x_3}| = |G_{x_4}| = 2$, and $|G_{x_5}| = 1$.

An *edge clique cover* of $G$ is any family $\mathcal{E} = \{Q_1, \ldots, Q_k\}$ of complete subgraphs of $G$ such that every edge of $G$ is in at least one of $E(Q_1), \ldots, E(Q_k)$; in other words, $xy \in E(G)$ implies $xy \in \cup_{i=1}^k E(Q_i)$. Remember that any of these $Q_i$'s may be the null subgraph of $G$. We customarily use $Q$'s (often with subscripts, superscripts, or other ornamentation) to denote complete subgraphs of $G$ or, interchangeably, the vertex sets of complete subgraphs.

Clearly, the set of all maxcliques of any graph $G$ forms an edge clique cover of $G$, as does the set $E(G)$ when each edge is viewed as a 2-element subset of $V(G)$. But a graph can have many other edge clique covers.

**Example 1.1 (continued)** For the given graph $G$, taking $Q_1 = G_{x_1}$ $= \{v_1, v_2, v_4\}$, $Q_2 = G_{x_2} = \{v_2\}$, $Q_3 = G_{x_3} = \{v_2, v_4\}$, $Q_4 = G_{x_4} = \{v_3, v_4\}$, and $Q_5 = G_{x_5} = \{v_4\}$ forms a 5-member edge clique cover $\mathcal{E}$ of $G$. Alternatively, $Q'_1 = \{v_1, v_2, v_4\}$, $Q'_2 = \{v_1, v_4\}$, $Q'_3 = \{v_3\}$, and $Q'_4 = \{v_3, v_4\}$ form a 4-member edge clique cover $\mathcal{Q}'$ of $G$.

The 5-member edge clique cover considered in Example 1.1 illustrates how each set representation $\mathcal{F} = \{S_1, \ldots, S_n\}$ of any intersection graph $G$ determines a *dual edge clique cover* $\mathcal{E}(\mathcal{F})$ of $G$ defined to be the family

$$\boxed{\mathcal{E}(\mathcal{F}) = \{G_x : x \in \cup_i S_i\}, \text{ where each } G_x = \{v_i : x \in S_i\},}$$

letting each $v_i \in V(G)$ correspond to $S_i \in \mathcal{F}$ under the isomorphism $G \cong \Omega(\mathcal{F})$.

Suppose $G$ is any graph with $V(G) = \{v_1, \ldots, v_n\}$. Every particular edge clique cover $\mathcal{E} = \{Q_1, \ldots, Q_k\}$ of $G$ determines a *dual set representation* $\mathcal{F}(\mathcal{E})$ of $G$ defined to be the family

$$\boxed{\mathcal{F}(\mathcal{E}) = \{S_1, \ldots, S_n\}, \text{ where each } S_i = \{j : v_i \in Q_j\}}$$

for each $i \in \{1, \ldots, n\}$. Observe that each $S_i$ in a dual set representation $\mathcal{F}(\mathcal{E})$ is a set of integers, and that $S_i \cap S_j \neq \emptyset$ if and only if $v_i v_j \in E(G)$.

**Example 1.1 (continued)** For the 5-member edge clique cover $\mathcal{E}(\mathcal{F})$ as above, the dual set representation $\mathcal{F}(\mathcal{E}(\mathcal{F}))$ consists of $S_1 = \{j : v_1 \in Q_j\} = \{1\}$, $S_2 = \{j : v_2 \in Q_j\} = \{1, 2, 3\}$, $S_3 = \{4\}$, and $S_4 = \{1, 3, 4, 5\}$. Notice how this set representation corresponds, set by set, to the $\mathcal{F}$ at the beginning of the example.

For the 4-member edge clique cover $\mathcal{E}' = \{Q'_1, Q'_2, Q'_3, Q'_4\}$ given earlier, the dual edge clique cover $\mathcal{F}(\mathcal{E}')$ consists of $S_1 = \{j : v_1 \in Q'_j\} = \{1, 2\}$, $S_2 = \{1\}$, $S_3 = \{3, 4\}$, and $S_4 = \{1, 2, 4\}$.

**Exercise 1.1** Given any graph $G$ with edge clique cover $\mathcal{E}$, show that the dual set representation $\mathcal{F} = \mathcal{F}(\mathcal{E})$ defined above actually *is* a set representation; in other words, show that $G \cong \Omega(\mathcal{F})$ with each $v_i \in V(G)$ corresponding to $S_i \in \mathcal{F}$.

**Exercise 1.2** Show that if $G$ is any intersection graph with set representation $\mathcal{F}$, then $\mathcal{F}(\mathcal{E}(\mathcal{F}))$ corresponds, set by set, to $\mathcal{F}$. Similarly, if $G$ is any intersection graph with edge clique cover $\mathcal{E}$, then $\mathcal{E}(\mathcal{F}(\mathcal{E}))$ corresponds, set by set, to $\mathcal{E}$.

The back-and-forth interplay—duality—between set representations and edge clique covers is a characteristic feature of intersection graph theory. We will see how it allows the interrelation of two different sorts of structures, each of which can be viewed as being represented by the other. Sections 1.4 and 1.5 will show examples of this, with many others appearing in later chapters. This interplay will show up in many of the results we present; it is a large part of what makes them work. (We present one enticing example in section 4.3: intersection graphs are used to consider whether ecological "food webs" can be represented by "competition graphs," and then whether those graph representations in turn have "interval representations"—back and forth and back again between set representations and edge clique covers.)

Every graph $G$ has the edge clique cover $\mathcal{E} = E(G)$, or at the other extreme $\mathcal{E}$ could consist of all the maxcliques of $G$. Thus Exercise 1.1 proves the "first theorem" of intersection graph theory, from [Marczewski, 1945].

**Theorem 1.1 (Marczewski)** *Every graph is an intersection graph.* □

While every graph has a set representation, intersection graph theory uses properties of the set representations and various conditions imposed thereon, rather than the conventional graph-theoretic properties that "forget" the sets. In many interesting examples a set representation $\mathcal{F}$ of a graph $G$ actually consists of the vertex sets of subgraphs of another graph $H$. We will often identify the vertex sets of subgraphs with the subgraphs themselves and say that $\mathcal{F}$ consists of the subgraphs. When this happens, we call $G$ the *guest graph*, $H$ the *host graph*, and the set representation a *graph representation* of $G$. Theorem 1.1 can be strengthened to show that every graph has a graph representation.

**Theorem 1.2** *Every graph $G$ is the intersection graph of a family of subgraphs of a graph.*

**Proof.** Suppose $G$ is any graph, $\mathcal{E} = \{Q_1, \ldots, Q_m\}$ is any edge clique cover of $G$, and $\mathcal{F} = \mathcal{F}(\mathcal{E})$ is the dual set representation of $G$ determined from $\mathcal{E}$; thus $G \cong \Omega(\mathcal{F})$. Define $H$ to have vertex set $\{1, \ldots, m\}$ with $ij \in E(H)$ if and only if $\{i, j\} \subseteq S_k$ for some $S_k \in \mathcal{F}$. Then each $S_k \in \mathcal{F}$ induces a complete subgraph of $H$ and, since $\mathcal{F}$ is a set representation of $G$, these induced complete subgraphs will form a graph representation of $G$. □

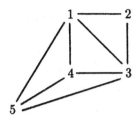
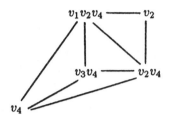

Figure 1.2: *A host graph $H$ for the graph $G$ in Figure 1.1 (and $H$ as a set-labeled intersection graph as in Lemma 1.3).*

**Example 1.1 (continued)** We illustrate the construction in the proof of Theorem 1.2 using the graph $G$ and the 5-member edge clique cover $\mathcal{E}$ given earlier in Example 1.1. The host graph $H$ corresponding to this guest graph $G$ is shown in Figure 1.2. The subgraph $S_1$ of $H$ is induced by $\{1\}$ since $Q_1$ is the only member of $\mathcal{E}$ that contains $v_1$, and $S_2$ is induced by $\{1, 2, 3\}$ since $Q_1$, $Q_2$, and $Q_3$ are the members of $\mathcal{E}$ that contain $v_2$; similarly, $S_3$ is induced by $\{4\}$ and $S_4$ by $\{1, 3, 4, 5\}$.

**Exercise 1.3** Suppose the pairs $G, \mathcal{E}$ and $H, \mathcal{F}$ are as in the proof of Theorem 1.2. Show that $\sum\{|Q_i| : Q_i \in \mathcal{E}\} = \sum\{|S_i| : S_i \in \mathcal{F}\}$.

Focusing on the graph $H$ constructed in the proof of Theorem 1.2, the following lemma shows how to go from a graph $G$ with edge clique cover $\mathcal{E}$ to a graph $H$ with edge clique cover $\mathcal{F}$ such that $G \cong \Omega(\mathcal{F})$ and $H \cong \Omega(\mathcal{E})$. Figure 1.2 shows $H$ from Example 1.1 as the set-labeled intersection graph $\Omega(\mathcal{E})$. Notice the symmetry—we can go either direction between $H, \mathcal{F}$ and $G, \mathcal{E}$, and so each graph can be thought of as a host for the other. We exploit this dual relationship between pairs of graphs in section 1.4 and later chapters.

**Lemma 1.3** *Suppose $G$ is any graph and $\mathcal{E} = \{Q_1, \dots, Q_m\}$ is any edge clique cover of $G$. Let $\mathcal{F}$ and $H$ be as in the proof of Theorem 1.2. Then $\mathcal{F}$ is an edge clique cover of $H$, $\mathcal{E} = \mathcal{E}(\mathcal{F})$, and $H \cong \Omega(\mathcal{E})$.*

**Proof.** This can be proved as a straightforward extension of the proof of Theorem 1.2, with each $i \in V(H)$ corresponding to $Q_i \in \mathcal{E}$ under the isomorphism $H \cong \Omega(\mathcal{E})$. □

**Exercise 1.4** Fill in the details in the proof of Lemma 1.3.

## 1.2    Intersection Classes

Theorem 1.1 shows that for every graph $G$ there is a family $\mathcal{F}$ of sets such that $G \cong \Omega(\mathcal{F})$. Interesting problems arise when restrictions are placed on $G$ and $\mathcal{F}$. Specifically, let $\mathcal{G}$ be a set of graphs and $\Sigma$ be a set of sets. We write $\mathcal{G} \cong \Omega(\Sigma)$ if each graph $G \in \mathcal{G}$ is isomorphic to an intersection graph $G' = \Omega(\mathcal{F})$ for some family $\mathcal{F}$ of sets from $\Sigma$ and, vice versa, each $G' = \Omega(\mathcal{F})$ for a family $\mathcal{F}$ from $\Sigma$ is isomorphic to a $G \in \mathcal{G}$. It is not always the case that each $\mathcal{G}$ has a $\Sigma$ for which $\mathcal{G} \cong \Omega(\Sigma)$, and the situation in which it does happen will be of considerable interest to us (for instance with $\Sigma$ the set of all subtrees of a tree in Chapter 2 and $\Sigma$ the set of all intervals of the real line in Chapter 3). Most of this section is based on [Scheinerman, 1985a], in which a set $\mathcal{G}$ of graphs is defined to be an *intersection class* if there is a $\Sigma$ such that $\mathcal{G} \cong \Omega(\Sigma)$.

A set $\mathcal{G}$ of graphs (or, equivalently, a property of graphs) is *closed under induced subgraphs* if $G' \in \mathcal{G}$ whenever $G'$ is an induced subgraph of some $G \in \mathcal{G}$. Equivalently, classes (properties) of graphs that are closed under induced subgraphs are precisely those that can be defined by a list of forbidden induced subgraphs—a potentially infinite list, with [McKee, 1978] describing what more is needed to ensure a finite list. As examples, the set of all planar graphs (or the graph-theoretic property of being planar) is closed under induced subgraphs, but the set of all connected graphs is not. The following exercise shows that the connected graphs do not form an intersection class.

**Exercise 1.5** Show that every intersection class is closed under induced subgraphs.

Define a set $\mathcal{G}$ of graphs to be *closed under vertex expansion* if $G' \in \mathcal{G}$ whenever $G'$ results from $G \in \mathcal{G}$ by repeatedly replacing an existing vertex $v$ by a pair $v', v''$ of new adjacent vertices, each having the same pre-existing neighbors as $v$ did. The set of all connected graphs is closed under vertex expansion, but the set of all planar graphs is not.

**Exercise 1.6** Show that every intersection class is closed under vertex expansion.

A set $\mathcal{G}$ of graphs has a *composition series* if there exists a countable sequence $\langle G_1, G_2, \ldots \rangle$ of graphs in $\mathcal{G}$ such that each $G_i$ is an induced subgraph of $G_{i+1}$ and each $G \in \mathcal{G}$ is the induced subgraph of some $G_i$. Notice that if the set $\mathcal{G}$ is closed under disjoint unions, then $\mathcal{G}$ has a composition series where, for instance, each $G_i$ can be taken to be the disjoint union of

$\{G \in \mathcal{G} : |V(G)| = i\}$. This shows that the set of all planar graphs has a composition series; somewhat similarly, so does the set of all connected graphs.

**Exercise 1.7 (Scheinerman)** Show that the set of all graphs that do not contain *both* cycles $C_4$ and $C_5$ as induced subgraphs does not have a composition series.

**Lemma 1.4 (Scheinerman)** *If $\mathcal{G}$ is an intersection class, then there is a countable $\Sigma$ such that $\mathcal{G} \cong \Omega(\Sigma)$, and $\mathcal{G}$ has a composition series.*

**Proof.** Consider an intersection class $\mathcal{G} \cong \Omega(\Sigma)$. Since there are only finitely many graphs of each possible order, $\mathcal{G}$ is certainly countable—say, $\mathcal{G} = \{G_1, G_2, \ldots\}$ where each $G_k \cong \Omega(\mathcal{F}_k)$ and each $\mathcal{F}_k \subseteq \Sigma$. Since each $V(G_k)$ is finite, for each $i$ there is a finite $\mathcal{F}'_k \subseteq \mathcal{F}_k$ such that $G_k \cong \Omega(\mathcal{F}'_k)$. Let $\Sigma'$ be the countable subset $\mathcal{F}'_1 \cup \mathcal{F}'_2 \cup \cdots$ of $\Sigma$. Then $\mathcal{G} \cong \Omega(\Sigma')$.

Therefore, we can assume that $\mathcal{G} \cong \Omega(\Sigma)$ where $\Sigma = \{S_1, S_2, \ldots\}$ is countable. Define graphs $H_1, H_2, \ldots$, where each $H_k$ has

$$V(H_k) = \{v_i^p : 1 \leq i \leq k \text{ and } 1 \leq p \leq k\}$$

and

$$E(H_k) = \{v_i^p v_j^q : (p,i) \neq (q,j) \text{ and } S_i \cap S_j \neq \emptyset\}.$$

Let each $\mathcal{F}''_k$ be the family consisting of $k$ copies of each of $S_1, \ldots, S_k$. Then making each $v_i^p$ correspond to $S_i$ produces an isomorphism $H_k \cong \Omega(\mathcal{F}''_k)$. Each $H_k$ is easily seen to be an induced subgraph of $H_{k+1}$. For each $G \in \mathcal{G}$, suppose $G \cong \Omega(\mathcal{F})$ where $\mathcal{F} \subseteq \Sigma$ and let $h$ be the maximum subscript $i$ for which a vertex of $G$ corresponds to an $S_i \in \mathcal{F}$ under that isomorphism. Setting $k = \max\{h, |V(G)|\}$ ensures that $G$ is an induced subgraph of $H_k$. Therefore, $\langle H_1, H_2, \ldots \rangle$ is a composition series for $\mathcal{G}$. □

**Theorem 1.5 (Scheinerman)** *A set $\mathcal{G}$ of graphs is an intersection class if and only if all three of the following conditions are satisfied:*

    *(1) $\mathcal{G}$ is closed under induced subgraphs;*

    *(2) $\mathcal{G}$ is closed under vertex expansion;*

    *(3) $\mathcal{G}$ has a composition series.*

*Moreover, if repeated members of $\Sigma$ are not allowed in the $\mathcal{F}$'s, then conditions (1) and (3) are necessary and sufficient.*

**Proof.** Exercises 1.5 and 1.6 and Lemma 1.4 prove the "only if" direction. For the "if" direction, suppose $\mathcal{G}$ satisfies conditions (1), (2), and (3) and has a composition series $\langle G_1, G_2, \ldots \rangle$. By (1) we can insert additional members into the composition series and, since each $G_i$ is an induced subgraph of $G_{i+1}$, we can even assume that each $V(G_i) = \{v_1, \ldots, v_i\}$. For each $i$, define

$$S_i = \{(j, i) : 0 < j < i \text{ and } v_i v_j \in E(G_i)\} \cup \{(i, j) : j > 0\}.$$

Let $\Sigma = \{S_1, S_2, \ldots\}$ (toward showing that $\mathcal{G} = \Omega(\Sigma)$).

Each $G_k \cong \Omega(\{S_1, \ldots, S_k\})$ since, for $i < j \leq k$, $S_i \cap S_j \neq \emptyset$ if and only if $S_i \cap S_j = \{(i, j)\}$; that is equivalent to $v_i v_j \in (G_j)$, and so to $v_i v_j \in (G_k)$. Each $G \in \mathcal{G}$ is, by condition (3), an induced subgraph of some $G_k$, and so $G \cong \Omega(\mathcal{F})$ such that $\mathcal{F} \subseteq \Sigma$ by Exercise 1.5. This shows one direction of $\mathcal{G} \cong \Omega(\Sigma)$.

Conversely, suppose $G \cong \Omega(\mathcal{F})$ where $\mathcal{F} \subseteq \Sigma$. Let $\mathcal{F}' = \{S_{i_1}, \ldots, S_{i_n}\}$ be the subset of $\mathcal{F}$ consisting of one copy of each distinct member of $\mathcal{F}$ (remember that the family $\mathcal{F}$ may have repeated members). Define a graph $G'$ on vertex set $\{w_1, \ldots, w_n\}$ where $w_p w_q \in E(G')$ if and only if $p \neq q$ and $S_p \cap S_q \neq \emptyset$. This makes $G'$ an induced subgraph of $G$, with $G$ resulting from $G'$ by vertex expansion. Since $G'$ is an induced subgraph of $G_k$ where $k = \max\{i_1, \ldots, i_n\}$, condition (1) implies that $G' \in \mathcal{G}$, and so condition (2) implies that $G \in \mathcal{G}$.

The "Moreover" portion of the theorem follows by a similar argument. $\square$

**Exercise 1.8** Fill in the details in the proof of the "Moreover" portion of the theorem, including checking Exercise 1.5 and Lemma 1.4 when the $\mathcal{F}$'s in $\Sigma$ are required to be *sets* rather than families.

While Scheinerman's theorem can be used to show that a particular set $\mathcal{G}$ is an intersection class, that is a long way from actually finding a suitable $\Sigma$ and proving that it works. For instance, Chapter 2 will define "chordal graphs" as graphs that have no induced cycles larger than triangles, and these graphs can easily be shown to satisfy all three conditions and so form an intersection class. Yet chordal graphs were studied for many years before an intersection characterization was found (or looked for); section 2.1 tells the story. As another example, planar graphs satisfy conditions (1) and (3)—but not condition (2)—and so always can be characterized as intersection graphs of families of distinct sets; yet in spite of this, no natural intersection characterization is known for them.

Scheinerman's approach is extended in [Scheinerman, 1985c, 1986], and [Quilliot, 1988] presents an abstract approach to similar questions in a hypergraph context.

[Moorhouse, 1994, to appear(a)] perform a similar analysis for *graph-based intersection classes*, the intersection graphs of families of *subgraphs* of a set $\Sigma$ of *graphs*. Perhaps surprisingly, this greater restriction on the objects being intersected allows less restriction on the graphs. Moorhouse shows that $\mathcal{G}$ is a graph-based intersection class if and only if $\mathcal{G}$ is closed under induced subgraphs and closed under vertex expansion. Moreover, if repeated members of $\Sigma$ are not allowed in the $\mathcal{F}$'s, then $\mathcal{G}$ being closed under induced subgraphs is necessary and sufficient. This work is extended in [Moorhouse, to appear(b)].

It should be noted that while Scheinerman's and Moorhouse's work gives very reasonable characterizations of those classes of graphs that are definable as intersection graphs, less stringent interpretations are possible. The following exercise, suggested only for those fond of arcana, contains an "intersection characterization" of hamiltonian graphs (a class of graphs that is not even closed under induced subgraphs!).

**Exercise 1.9** (see [Zamfirescu, 1973/74]) Show that a graph $G$ is hamiltonian if and only if there exists a family $\mathcal{F} = \{C_1, \ldots, C_n\}$ of cycles of $G$ such that the following three conditions hold:

- every vertex of $G$ is in at least one cycle in $\mathcal{F}$;
- the intersection-like graph $F^*$ is a tree, where $F^*$ is defined to have $V(F^*) = \mathcal{F}$ with $C_i C_j \in E(F^*)$ if and only if the subgraph $C_i \cap C_j$ consists precisely of a single edge; and
- the intersection graph $\Omega(\mathcal{F})$ is a tree, where each $C_i$ is now viewed as a subset of $V(G)$.

## 1.3 Parsimonious Set Representations

Since every graph is an intersection graph, it may seem that more structure has to be required of the set representation in order to ask interesting questions about particular graphs. But several challenging problems arise instantly, including finding smallest set representations and identifying when a set representation is unique. Define the *intersection number* $i(G)$ to be the minimum cardinality of a set $S$ such that $G$ is an intersection graph of a family of subsets of $S$.

**Exercise 1.10** Show that $i(K_2) = 1$, $i(P_3) = 2$, $i(2K_2) = 2$, and $i(K_3) = 1$.

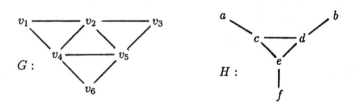

Figure 1.3: *A graph G with intersection number 3. (Graph H will be explained in Section 1.5.)*

Our next result characterizes $i(G)$ in terms of the more "internal" parameter $\theta(G)$, the minimum cardinality of an edge clique cover of $G$. Theorem 1.6 was proved in [Erdős, Goodman, & Pósa, 1966] and has been rediscovered several times by other authors in slightly different contexts.

**Theorem 1.6 (Erdős, Goodman, & Pósa)** *For every graph $G$, $i(G)$ $= \theta(G)$.*

**Proof.** Let $\mathcal{E}$ be an edge clique cover of $G$ with $|\mathcal{E}| = \theta(G)$. Then the set representation $\mathcal{F} = \mathcal{F}(\mathcal{E})$ of $G$ has $|\cup \{S_i : S_i \in \mathcal{F}\}| = \theta(G)$, so that $i(G) \leq \theta(G)$. Conversely, since $G$ has a set representation by Theorem 1.2, we can pick $\mathcal{F}$ to have $|\cup \{S_i : S_i \in \mathcal{F}\}|$ minimum. Then $\mathcal{F}$ determines the edge clique cover $\mathcal{E} = \mathcal{E}(\mathcal{F})$ of $G$ with $|\mathcal{E}| = |\cup \{S_i : S_i \in \mathcal{F}\}| = i(G)$, so that $\theta(G) \leq i(G)$.  □

**Example 1.2** If $G$ is as in Figure 1.3, then $\theta(G) = 3$: taking $\mathcal{E}$ to consist of $Q_1 = \{v_1, v_2, v_4\}$, $Q_2 = \{v_2, v_3, v_5\}$, and $Q_3 = \{v_4, v_5, v_6\}$ shows that a 3-member edge clique cover is sufficient, and it is easy to see that no fewer than three will work. Observe that $\mathcal{F}(\mathcal{E}) = \{\{1\}, \{1,2\}, \{2\}, \{1,3\}, \{2,3\}, \{3\}\}$ is a set representation of $G$ with $\cup\{S_i : S_i \in \mathcal{F}\}$ of minimum cardinality.

It is not easy in general to determine $\theta(G)$ or $i(G)$—in fact [Kou, Stockmeyer, & Wong, 1978] shows it to be NP-hard—but they have been determined for some special cases. Recall that a *triangle-free graph* is a graph that does not contain $K_3$ as a subgraph.

**Corollary 1.7** *Every graph $G$ has $i(G) \leq |E(G)|$, with $i(G) = |E(G)|$ if and only if $G$ is triangle-free.*  □

Now define $i^*(G)$ to be the minimum cardinality of a set $S$ such that $G$ is an intersection graph of *distinct* subsets of $S$. Clearly $i(G) \leq i^*(G)$ for every graph $G$. (Warning: Some authors call $i^*(G)$ the "intersection number" and $i(G)$ the "pseudointersection number" of $G$.)

**Exercise 1.11** Show $i^*(K_2) = 2$, $i^*(P_3) = 2$, $i^*(2K_2) = 4$, and $i^*(K_3) = 3$.

**Exercise 1.12** Modify the proof of Theorem 1.1 to show that every graph is the intersection graph of a family of distinct sets.

If $v \in V(G)$, then the *closed neighborhood of* $v$, denoted $N[v]$, is the set of all vertices of $G$ adjacent to $v$ together with $v$ itself. A graph $G$ is *point determining* if, for all $u, v \in V(G)$ with $u \neq v$, $N[u] \neq N[v]$. [Sumner, 1973] introduced this notion, and [Lim, 1978], calling them *supercompact graphs*, contains many characterizations and properties.

**Exercise 1.13** (see [Slater, 1976]) Show that if $G$ is a point determining graph with no isolated vertices, then $i(G) = i^*(G)$.

**Corollary 1.8** *If $G$ is triangle-free and each component has at least three vertices, then $i(G) = i^*(G)$.*                    $\square$

**Corollary 1.9** *If $G$ is a connected graph with $|V(G)| \geq 4$, then $i^*(G) = |E(G)|$ if and only if $G$ is triangle-free.*                    $\square$

**Exercise 1.14** Show that the converse to Exercise 1.13 is not true.

Note that if $G$ is a triangle, then $i^*(G) = 3 = |E(G)|$, so the hypothesis of $|V(G)| \geq 4$ is necessary in Corollary 1.9.

**Theorem 1.10** (see [Erdős, Goodman, & Pósa, 1966]) *For any graph $G$ with $n = |V(G)|$, $i(G) \leq \lfloor n^2/4 \rfloor$.*

**Proof.** First note that we may assume that $G$ contains no isolated vertices. We show the stronger result that there is an edge clique cover of $G$ that consists of at most $\lfloor n^2/4 \rfloor$ edges and triangles of $G$.

The result is easily checked for $n = 2, 3$. By way of induction, assume the result is true for all graphs that have no more than $n + 2$ vertices, and suppose $|V(G)| = n+2$. Pick $xy \in E(G)$ and consider the graph $G' = G \backslash \{x, y\}$. By the inductive hypothesis, $G'$ has an edge clique cover that consists of at most $\lfloor n^2/4 \rfloor$ edges and triangles. By considering for each $v \in V(G')$ whether the subgraph induced by $\{x, y, v\}$ is $K_3$, $P_3$, or $K_1 \cup K_2$, it is clear that at most $n + 1$ additional edges or triangles are needed to make an edge clique cover of $G$. Since $\lfloor (n+2)^2/4 \rfloor = \lfloor n^2/4 \rfloor + n + 1$, the proof is complete.  $\square$

A slightly different proof technique than above shows that, for any graph $G$ with $n = |V(G)| \geq 4$, $i^*(G) \leq \lfloor n^2/4 \rfloor$.

**Exercise 1.15** Show that the number $\lfloor n^2/4 \rfloor$ is best possible in Theorem 1.10 by finding a graph of order $n$ that requires $\lfloor n^2/4 \rfloor$ members in every edge clique cover.

We now turn briefly to the question of uniqueness. Let $G$ be a graph that is an intersection graph of a family of distinct subsets of $S$ where $|S| = i^*(G)$. Then $G$ is said to be *uniquely intersectable* if, for every two families $\mathcal{F}_1$ and $\mathcal{F}_2$ of *distinct* subsets of $S$, $\Omega(\mathcal{F}_1) \cong \Omega(\mathcal{F}_2) \cong G$ implies that $\mathcal{F}_1$ can be obtained from $\mathcal{F}_2$ by permuting the elements of $S$.

**Example 1.3** The cycle $C_4$ is uniquely intersectable since $i^*(C_4) = 4$ and, for each $x \in S$ where $|S| = 4$, $x$ is in exactly two sets in any $\mathcal{F}$, with four distinct subsets required.

The complete graph $K_3$ is not uniquely intersectable. To see this, first note that $i^*(G) = 3$. Let $S = \{a, b, c\}$. Now, $\Omega(\mathcal{F}_1) \cong \Omega(\mathcal{F}_2) \cong G$, where $\mathcal{F}_1 = \{\{a, b\}, \{a, c\}, \{b, c\}\}$ and $\mathcal{F}_2 = \{\{a\}, \{a, b\}, \{a, b, c\}\}$. Clearly $\mathcal{F}_1$ cannot be obtained from $\mathcal{F}_2$ by permuting the elements of $S$.

Corollary 1.9 shows that the condition of being triangle-free can lead to a nice result. The following is another example of this.

**Exercise 1.16** (see [Alter & Wang, 1977]) Show that every triangle-free graph is uniquely intersectable.

Alter and Wang also show that no $K_n$ with $n \geq 3$ is uniquely intersectable and give many types of uniquely intersectable graphs. However, the problem of giving a complete characterization of uniquely intersectable graphs remains open. [Mahadev & Wang, 1997, to appear] contains more recent developments. [Era & Tsuchiya, 1991] and [Tsuchiya, 1994] discuss intersection numbers when conditions are placed on the family $\mathcal{F}$ of subsets of $S$, for instance when $\mathcal{F}$ is an *antichain*, meaning that no two members of $\mathcal{F}$ are comparable.

[Lim & Peng, 1991] defines *uniquely pseudointersectable graphs* by dropping the requirement that the members of $\mathcal{F}_1$ and $\mathcal{F}_2$ be distinct in the definition of uniquely intersectable graphs and shows that the notions of uniquely intersectable and unique pseudointersectable are equivalent for point determining graphs.

**Exercise 1.17** Show that the complete graph $K_3$ is uniquely pseudointersectable.

## 1.4 Clique Graphs

Recall that a maxclique of a graph is a complete subgraph that is not properly contained in another complete subgraph.

**Exercise 1.18** Given maxcliques $Q$ and $Q'$ of $G$ with $v \in Q$ such that $v \notin Q'$, show that there exists $v' \in Q'$ such that $v' \notin Q$ and $vv' \notin E(G)$.

We define the *clique graph operator* $K(\cdot)$ such that, for any graph $H$, $K(H)$ is the intersection graph of *all* the maxcliques of $H$. A graph is a *clique graph* if it is isomorphic to $K(H)$ for some graph $H$.

Clique graphs (and the clique graph operator) will be important to us in later chapters. They are characterized in [Roberts & Spencer, 1971] in terms of the following condition. A family $\mathcal{F} = \{S_1, \ldots, S_k\}$ of subsets of a set $S$ is said to satisfy the *Helly condition* if the following holds: For every subfamily $\mathcal{F}' \subseteq \mathcal{F}$, if the members of $\mathcal{F}'$ intersect pairwise, then all the members have a common element—in other words, if every $S_i, S_j \in \mathcal{F}'$ has $S_i \cap S_j \neq \emptyset$, then $\cap\{S_i : S_i \in \mathcal{F}'\} \neq \emptyset$.

**Lemma 1.11** *Suppose a graph $G$ has edge clique cover $\mathcal{E} = \{Q_1, \ldots, Q_m\}$ determining the dual set representation $\mathcal{F} = \mathcal{F}(\mathcal{E})$ of $G$. Define a graph $H$ on $V(H) = \{1, \ldots, m\}$ such that $H \cong \Omega(\mathcal{E})$ with each $i \in V(H)$ corresponding to $Q_i$ under that isomorphism. Then $\mathcal{E}$ satisfies the Helly condition if and only if $\mathcal{F}$ contains every maxclique of $H$.*

**Proof.** Suppose $G$, $\mathcal{E}$, $\mathcal{F}$, and $H$ are as in the statement of the lemma. By Lemma 1.3, $\mathcal{F}$ is an edge clique cover of $H$, making each $S_i \in \mathcal{F}$ induce a complete subgraph of $H$, and $H \cong \Omega(\mathcal{E})$.

Suppose $\mathcal{E}$ satisfies the Helly condition and $R$ is any maxclique of $H$ (toward showing that $R \in \mathcal{F}$). If $j, k \in R$, then $jk \in E(H)$ and so $Q_j \cap Q_k \neq \emptyset$ by $H \cong \Omega(\mathcal{E})$; thus the subfamily $\{Q_j : j \in R\}$ of $\mathcal{E}$ has pairwise nonempty intersections. By the Helly condition, there is some $v_i \in \cap\{Q_j : j \in R\}$. So $j \in R$ implies $v_i \in Q_j$, which implies $j \in S_i = \{j : v_i \in Q_j\}$ by the definition of $\mathcal{F}(\mathcal{E})$. Thus $R \subseteq S_i$ and so, since $R$ is a maxclique, $R = S_i \in \mathcal{F}$.

Conversely, suppose $\mathcal{F} = \mathcal{F}(\mathcal{E}) = \{S_1, \ldots, S_n\}$ contains every maxclique of $H$ and we are given $\mathcal{E}' \subseteq \mathcal{E}$ that has pairwise nonempty intersections (toward showing that some $v_i \in V(G)$ is in every $Q_j \in \mathcal{E}'$). Thus, for every $Q_j, Q_k \in \mathcal{E}'$ there is some $v_i \in Q_j \cap Q_k$, and so $j, k \in S_i = \{j : v_i \in Q_j\}$ by the definition of $\mathcal{F}(\mathcal{E})$; since $S_i$ induces a complete subgraph of $H$, this implies $jk \in E(H)$. Thus $\{j : Q_j \in \mathcal{E}'\}$ induces a complete subgraph of $H$ and so is contained in some $S_i \in \mathcal{F}$ that is a maxclique of $H$. By the definition of $\mathcal{F}(\mathcal{E})$, $v_i$ is then contained in every $Q_j \in \mathcal{E}'$.  □

Notice that the final conclusion on the dual set representation $\mathcal{F}$ in Lemma 1.11 can be restated as follows: For every subset $V' \subseteq V(H)$, if every two elements of $V'$ are in a common member of $\mathcal{F}$, then all the elements of $V'$ are in a common member of $\mathcal{F}$. This situation is sometimes described as $\mathcal{F}$ satisfying the *conformality condition*, dual to the Helly condition.

**Theorem 1.12 (Roberts & Spencer)** *A graph is a clique graph if and only if it has an edge clique cover that satisfies the Helly condition.*

**Proof.** Given any graph $H \cong K(G)$, let $\mathcal{E}$ be the edge clique cover of $G$ consisting of the maxcliques of $G$; thus $H \cong \Omega(\mathcal{E})$. Then $\mathcal{F} = \mathcal{F}(\mathcal{E})$ is an edge clique cover of $H$ by Lemma 1.3 and satisfies the Helly condition by Lemma 1.11.

Conversely, suppose a graph $G$ has an edge clique cover $\mathcal{E}$ that satisfies the Helly condition. Let $H$ and $\mathcal{F} = \{S_1, \ldots, S_n\}$ be as in Lemma 1.3, so $G = \Omega(\mathcal{F})$. Define $H^*$ on $V(H^*) = V(H) \cup \mathcal{F}$ to have $E(H^*) = E(H) \cup \{jS_i : j \in S_i\}$. Each $S_i \in \mathcal{F}$ is a vertex of $H^*$ that is in a unique maxclique of $H^*$—namely, $S_i \cup \{S_i\}$. Since $\mathcal{F}$ contains every maxclique of $H$ by Lemma 1.11, each maxclique of $H^*$ contains a unique vertex $S_i \in \mathcal{F}$. Thus $G \cong \Omega(\mathcal{F})$ ensures that $G \cong K(H^*)$, showing that $G$ is indeed a clique graph.  □

**Exercise 1.19** Use the proof of Theorem 1.12 to find an $H$ such that $K(H)$ is the graph in Figure 1.1. Repeat for the graph in Figure 7.12.

**Exercise 1.20** Show that the graph $G$ in Figure 1.3 is not a clique graph.

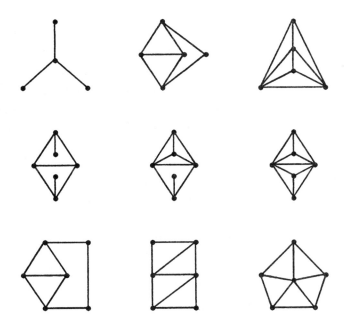

Figure 1.4: *The graphs that cannot be induced subgraphs of line graphs.*

## 1.5  Line Graphs

We include a discussion of line graphs since they form the first intersection class to be widely studied and since they typify much that is common to all intersection classes. They are also somewhat opposite in nature to clique graphs, which are based on maxcliques, in that line graphs are based on edges, which could be called "mincliques."

We define a *line graph operator* $L(\cdot)$ such that, for any graph $H$, $L(H)$ is the intersection graph of *all* the edges of $H$, each viewed as a 2-element subset of $V(H)$. A graph is a *line graph* if it is isomorphic to $L(H)$ for some graph $H$. [Hemminger & Beineke, 1978] surveys the extensive literature on line graphs up to that date, and [Prisner, 1996a] discusses many more recent results and generalizations.

**Example 1.4** Show that $L(K_3) \cong L(K_{1,3})$ ($K_{1,3}$ is the upper-left graph in Figure 1.4). [Whitney, 1932] shows that these are the only two nontrivial graphs that have isomorphic line graphs.

The following theorem, from [Krausz, 1943], is the prototype of what are sometimes called *Krausz-type characterizations*, meaning the characteriza-

tion of an intersection class by requiring each graph of that class to possess a
family of complete subgraphs that satisfies some sort of property intimately
related to the specific intersection class being studied. This is obviously only
a rough description; we give several examples in this monograph; and [Mc-
Kee, 1991a] contains formal details and shows a sense in which Krausz-type
characterizations can be, typically in hindsight, mechanically constructed
from the intersection definitions. This analysis requires formulating graph-
theoretic properties within a formal logical system. This is similar to what
is done in [McKee, 1991d] for certain characterizations of chordal graphs
(Chapter 2) and of interval graphs (Chapter 3), for instance showing how
their intersection definitions can lead, again in hindsight, to other characterizations.

The following theorem can be thought of as translating the properties
"every edge has exactly two vertices," and "no two edges have two vertices in
common"—in other words, no loops or parallel edges—into simple conditions
on an edge clique cover. This sort of translation is common to many of our
theorems; Lemma 1.11 can also be viewed as an example, translating "every
complete subgraph is contained in a maxclique" into the Helly condition on
an edge clique cover.

**Theorem 1.13 (Krausz)** *A graph $G$ is a line graph if and only if it
has an edge clique cover $\mathcal{E}$ such that both the following conditions hold:*
   *(1) every vertex of $G$ is in exactly two members of $\mathcal{E}$;*
   *(2) every edge of $G$ is in exactly one member of $\mathcal{E}$.*

**Proof.** First suppose $G \cong L(H)$ and let $\mathcal{F}$ be the edge clique cover of
$H$ that consists of the edges of $H$. Thus $G \cong \Omega(\mathcal{F})$, and we can suppose
subscripts are assigned so that each $v_i \in V(G)$ corresponds to $S_i \in \mathcal{F}$
under that isomorphism. Let $\mathcal{E} = \mathcal{E}(\mathcal{F})$ be the dual edge clique cover of $G$
determined from $\mathcal{F}$. Then for each $v_i \in V(G)$, $|\{x : v_i \in G_x\}| = |\{x : x \in
S_i\}| = |S_i| = 2$, and so condition (1) holds. Similarly for each $v_i v_j \in E(G)$,
$|\{x : v_i, v_j \in G_x\}| = |\{x : x \in S_i, S_j\}| \leq 1$ and equals 1 since $\mathcal{E}$ is an edge
clique cover, and so condition (2) holds.

Conversely, suppose $G$ has an edge clique cover $\mathcal{E} = \{Q_1, \ldots, Q_m\}$ that
satisfies conditions (1) and (2). Let $H = \Omega(\mathcal{E})$ and let $\mathcal{F} = \mathcal{F}(\mathcal{E})$ be the dual
set representation of $G$. Thus $G \cong \Omega(\mathcal{F})$, and we can suppose subscripts
are assigned so that each $S_i \in \mathcal{F}$ corresponds to $v_i \in V(G)$ under that
isomorphism. For each edge $Q_j Q_k$ of $H$ there exists some $v_i \in Q_j \cap Q_k$,
and so some $S_i$ exists that contains both $j$ and $k$. By condition (1), each
$|S_i| = |\{j : v_i \in Q_j\}| = 2$, and so each $Q_j Q_k \in E(H)$ corresponds to

$S_i = \{j, k\} \in \mathcal{F}$. Moreover, by condition (2), each $|S_i \cap S_k| = |\{j : v_i, v_k \in Q_j\}| \le 1$, so the members of $\mathcal{F}$ are distinct. Therefore, $\mathcal{F} = E(H)$, and so $G \cong \Omega(\mathcal{F}) \cong L(H)$, showing that $G$ is indeed a line graph.     $\square$

**Example 1.5** In Figure 1.3, $G$ is the line graph of $H$. In the first part of the proof of the theorem, taking $S_1 = \{a, c\}$, $S_2 = \{c, d\}$, $S_3 = \{b, d\}$, $S_4 = \{c, e\}$, $S_5 = \{d, e\}$, and $S_6 = \{e, f\}$ for $\mathcal{F}$ leads to $G_a = \{v_1\}$, $G_b = \{v_3\}$, $G_c = \{v_1, v_2, v_4\}$, and so on for $\mathcal{E}$. In proving the converse, taking $Q_1 = \{v_1, v_2, v_4\}$, $Q_2 = \{v_2, v_3, v_5\}$, $Q_3 = \{v_4, v_5, v_6\}$, $Q_4 = \{v_1\}$, $Q_5 = \{v_3\}$, and $Q_6 = \{v_6\}$ for $\mathcal{E}$ leads to $S_1 = \{1, 4\}$, $S_2 = \{1, 2\}$, and so on for $\mathcal{F}$.

**Exercise 1.21** Use the proof of Theorem 1.13 to find an $H$ such that $L(H)$ is the graph in Figure 1.1.

**Exercise 1.22** Choose any three graphs in Figure 1.4 and show that they are not line graphs.

Unlike for clique graphs, other characterizations are available for line graphs that do not involve finding edge clique covers. For instance, [Beineke, 1968] shows that a graph is a line graph if and only if it has none of the graphs in Figure 1.4 as an induced subgraph. Efficient recognition algorithms appear in [Roussopoulos, 1973] and [Lehot, 1974].

Line graphs can be generalized to many other sorts of intersection graphs, for instance using the intersection of other kinds of induced subgraphs (instead of edges—those subgraphs isomorphic to $K_2$), where each is viewed as a set of still other kinds of induced subgraphs (instead of vertices—those subgraphs isomorphic to $K_1$). [Cai, Corneil, & Proskurowski, 1996] discusses such generalizations.

## 1.6  Hypergraphs

Many of the concepts of intersection graph theory have natural analogues for hypergraphs—indeed, they have frequently been developed within hypergraph theory. Because of that, we include sections on hypergraphs in the first three chapters, introducing terminology as needed; hypergraphs also appear throughout Chapter 7. The present section shows how hypergraphs interconnect the ideas from earlier in the present chapter.

A *hypergraph* $H = (X, \mathcal{E})$ consists of a finite set $X$ of *vertices* and a family $\mathcal{E} = \{S_1, \ldots, S_n\}$ of *edges*—nonempty subsets of $X$. [Berge, 1989] is a standard reference for hypergraph theory, although we warn the reader that terminology and notation are far from standardized. [Duchet, 1995] is a recent thorough survey.

A hypergraph $(X, \mathcal{E})$ is a *simple hypergraph* when the family $\mathcal{E}$ is a *set*, that is, when all the edges are distinct. Thus, graphs are precisely the simple hypergraphs in which each edge contains exactly two vertices. A hypergraph $(X, \mathcal{E})$ is a *Helly hypergraph* when $\mathcal{E}$ satisfies the Helly condition from section 1.4. Because Helly hypergraphs will be very important to us later, we include the following useful *Gilmore criterion* from [Roberts & Spencer, 1971].

**Exercise 1.23 (Berge & Gilmore)** Show that a hypergraph $(X, \mathcal{E})$ is a Helly hypergraph if and only if, for every $u, v, w \in X$, there exists $x \in X$ such that every edge in $\mathcal{E}$ that contains at least two of $u, v, w$ also contains $x$. (Hint: Use induction on $|\mathcal{E}'|$, $\mathcal{E}' \subseteq \mathcal{E}$, for the harder direction.)

The *line graph* of the hypergraph $(X, \mathcal{E})$ is defined to be $\Omega(\mathcal{E})$. Theorem 1.2 implies that every graph is isomorphic to the line graph of a hypergraph, but the following theorem shows that more is true.

**Theorem 1.14** *Every graph is isomorphic to the line graph of a Helly hypergraph.*

**Proof.** Suppose $G$ is any graph and $\mathcal{E} = \{Q_1, \ldots, Q_m\}$ is the edge clique cover of $G$ consisting of the maxcliques of $G$. Let $\mathcal{F} = \mathcal{F}(\mathcal{E})$ be the set representation of $G$ determined from $\mathcal{E}$, and let $H$ be the hypergraph $(\{1, \ldots, m\}, \mathcal{F})$. Then $G \cong \Omega(\mathcal{F})$ implies that $G \cong L(H)$, and $H$ can be shown to be a Helly hypergraph.                                                    $\square$

**Exercise 1.24** Finish the proof of the preceding theorem by showing that $\mathcal{F}$ satisfies the Helly condition.

# Chapter 2

# Chordal Graphs

A graph is a *chordal graph* if it has no induced cycles larger than triangles. A *chord of a cycle* is an edge between nonconsecutive vertices of the cycle; thus a graph is chordal if and only if every cycle large enough to have a chord does have a chord. The study of chordal graphs goes back to [Hajnal & Surányi, 1958], frequently under the names *rigid-circuit graphs* or *triangulated graphs*. Chapter 4 of [Golumbic, 1980] is the standard reference for chordal graphs. [Blair & Peyton, 1993] is more up to date and more in the style presented here.

In spite of there having been considerable activity during the 1960s, it was not until the 1970s that chordal graphs were characterized in terms of intersection graphs. Many of the most sophisticated applications of chordal graphs, which we sketch in section 2.4, came later and involved the rediscovery of chordal graph theory in statistics and matrix analysis. The recent dates on many of our references show that chordal graphs are still being intensively studied today.

Contrary to history, we begin with the intersection graph approach to chordal graphs.

## 2.1 Chordal Graphs as Intersection Graphs

For the purpose of this section only, we define a graph to be a *subtree graph* if it is the intersection graph of a family of subtrees of a tree. But you should keep in mind that Theorem 2.4 at the end of this section will show that *the subtree graphs are precisely the chordal graphs*! The tree and family of subtrees in the definition are called a *tree representation* of the subtree graph and, while a tree is a topological object, it is clear that it can always

19

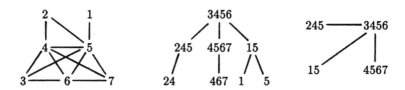

Figure 2.1: *A chordal graph and two tree representations.*

be taken to be a tree in the graph-theoretic sense.

**Example 2.1** The graph $G$ shown on the left in Figure 2.1 is a subtree graph isomorphic to $\Omega(\{T_1, \ldots, T_7\})$ where each $T_i$ is the subtree of the tree in the middle induced by those vertices that contain $i$. For instance, $V(T_5) = \{15, 245, 3456, 4567, 5\}$. There are, of course, many such tree representations of $G$. For instance, the tree shown on the right is a tree representation for $G$, but now the vertex set is precisely the set of maxcliques of $G$.

It is easy to see that $G$ is a subtree graph if and only if it has an edge clique cover $\mathcal{E}$ whose members can be associated with vertices of a tree $T$ such that, for every $v \in V(G)$, $\{Q : v \in Q \in \mathcal{E}\}$ induces a subtree $T_v$ of $T$. This is a very transparent translation of being a subtree graph into a condition on an edge clique cover. Theorem 2.1 shows that the edge clique cover can always be taken to be the set of the maxcliques of $G$. Theorem 2.3 then shows how to test whether the maxcliques of $G$ can be arranged into a tree as just described.

When a tree representation exists whose vertex set is the set of max-cliques of $G$, then it is called a *clique tree representation* (or a *clique tree* for) $G$. Equivalently, a clique tree is a spanning tree of the clique graph $K(G)$ such that, for each $v \in V(G)$, $T_v$ is connected. (Lemma 2.2 will give an alternative condition to check.) Given any clique tree $T$ for $G$ and any two maxcliques $Q_i$ and $Q_j$ of $G$, let $T(Q_i, Q_j)$ denote the path in $T$ connecting $Q_i$ and $Q_j$.

**Exercise 2.1** Show that in any clique tree $T$ for a chordal graph $G$, the family $\{T_v : v \in V(G)\}$ of subtrees of $T$, with each subtree viewed as a set of vertices of $T$, satisfies the Helly condition.

**Theorem 2.1** *A graph is a subtree graph if and only if it has a clique tree representation.* □

**Exercise 2.2** Use Lemma 1.11 to prove Theorem 2.1.

**Exercise 2.3** Show that every subtree graph is the intersection graph of a family of *distinct* subtrees of a tree. Is every subtree graph the intersection graph of a family of distinct subtrees of a *clique* tree?

The following lemma essentially appears in [Acharya & Las Vergnas, 1982] (see also [Levin, 1983]) modulo knowing other results that we prove in this section and the next; the lemma seems to first appear in this simple "clique tree check" form in [McKee, 1993].

**Lemma 2.2** *A spanning subtree $T$ of $K(G)$ is a clique tree for a connected graph $G$ if and only if*

$$|V(G)| = \sum_{Q \in V(T)} |Q| - \sum_{Q_i Q_j \in E(T)} |Q_i \cap Q_j|. \qquad (2.1)$$

**Proof.** Suppose $T$ is a spanning tree of $K(G)$. For each $v \in V(G)$, the subgraph $T_v$ satisfies $1 \le |V(T_v)| - |E(T_v)|$, with equality if and only if $T_v$ is connected and thus is a subtree. Summing over all $v \in V(G)$ proves equality (2.1). □

**Example 2.2** The cycle $C_4$ is not a subtree graph: each of the four spanning trees of $K(C_4)$ leaves one $T_v$ disconnected, and $4 < 8 - 3$ in equality (2.1).

**Exercise 2.4** Show that a subtree graph of order $n$ can have at most $n$ maxcliques.

Theorem 2.3 will show how easy it is to find clique tree representations of subtree graphs. It first appeared in [Bernstein & Goodman, 1981] in the computer science context we discuss in section 2.4, and it has been rediscovered many times. [Gavril, 1987] and [Shibata, 1988] give nice treatments.

It is important to realize that the approach in Theorem 2.3 requires knowing all the maxcliques of $G$, a computationally hard problem in general—the number of maxcliques of $G$ can grow exponentially in the number of vertices of $G$—yet one that can be done efficiently for subtree graphs because of Exercise 2.4. In certain applications, for instance the one in subsection 2.4.4 below, $G$ is given at the start as the set of its maxcliques.

**Exercise 2.5** Show that the order-$2p$ complete $p$-partite graph $K_{2,\ldots,2}$ has $2^p$ maxcliques.

For any graph $G$, define the *weighted clique graph* $K^w(G)$ to be the clique graph $K(G)$ with each edge $Q_iQ_j$ given weight $|Q_i \cap Q_j|$. Theorem 2.3 will involve maximum spanning trees of $K^w(G)$, which can be found efficiently by using Kruskal's well-known greedy algorithm. Recall that the usual *minimum* spanning tree version of Kruskal's algorithm finds a(ll) minimum spanning tree(s) of a connected weighted graph by repeatedly choosing an edge of smallest weight that does not form a cycle with previously chosen edges. The *maximum* spanning tree version that we use is the same, except that we now always choose an edge with *largest* weight that does not form a cycle with previously chosen edges.

**Example 2.3** For the graph $G$ on the left in Figure 2.1, a maximum spanning tree of $K^w(G)$ must contain the weight-three edge joining vertex 3456 to 4567, one of the two weight-two edges incident with 245, and one of the three weight-one edges incident with 15; one maximum spanning tree is shown on the right in the figure. Checking that such a tree is a clique tree requires either checking that each of the seven $T_i$'s is connected or checking that $7 = 21 - 14$ in equality (2.1).

**Theorem 2.3** *A connected graph $G$ is a subtree graph if and only if some maximum spanning tree of $K^w(G)$ is a clique tree for $G$. Moreover, this is equivalent to every maximum spanning tree of $K^w(G)$ being a clique tree for $G$, and every clique tree of $G$ is such a maximum spanning tree.*

**Proof.** If some maximum spanning tree of $K^w(G)$ is a clique tree for $G$, then by definition $G$ is a connected subtree graph.

Conversely, suppose $G$ is a connected subtree graph with clique tree $T$. Thus $T$ is a spanning tree of $K^w(G)$, but suppose, arguing toward a contradiction, that $T$ is not a *maximum* spanning tree of $K^w(G)$. Among all maximum spanning trees of $K^w(G)$, choose $T'$ to have a maximum number of edges in common with $T$. Pick any edge $e = Q_iQ_j \in E(T') \setminus E(T)$ having weight $|Q_i \cap Q_j|$ as large as possible. Since $T$ is a tree representation of $G$, each $v \in V(G)$ that is in $Q_i \cap Q_j$ must also be in every vertex of the path $T(Q_i, Q_j)$ in $T_v$, and so each edge of this path must have weight at least $|Q_i \cap Q_j|$. There must be some edge $f$ of this path that is not in $E(T')$. But the spanning tree $T'' = T' - e + f$ then has total weight at least as large as the weight of $T'$. Thus $T''$ is a maximum spanning tree of $K^w(G)$ having

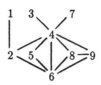

Figure 2.2: *A graph having three minimal vertex separators.*

one more edge in common with $T$ than does $T'$, contradicting the choice of $T'$.

Therefore, being a clique tree implies some maximum spanning tree of $K^w(G)$ is a clique tree. The rest of the theorem follows from Lemma 2.2 since every maximum spanning tree $T$ of $K^w(G)$ will have the same total weight $\sum_{QQ' \in E(T)} |Q \cap Q'|$.                                     $\square$

A set $S$ of vertices of $G$ is a *minimal vertex separator* of $G$ whenever there exist $u, v \in V(G)$ such that every path connecting $u$ and $v$ contains a vertex in $S$ and no proper subset of $S$ has this same property.

**Example 2.4** In the graph shown in Figure 2.2, the minimal vertex separators are $\{2\}$, $\{4\}$, and $\{4, 6\}$.

The following two exercises show how Kruskal's algorithm locates the minimal vertex separators of a subtree graph and that, even though a subtree graph can have many clique trees $T$, the multiset $\{Q_i \cap Q_j : Q_i Q_j \in E(T)\}$ is uniquely determined.

**Exercise 2.6** (see [Barrett, Johnson, & Lindquist, 1989] and [Ho & Lee, 1989]) Suppose $G$ is a connected subtree graph with clique tree $T$ and $S \subseteq V(G)$. Show that $S$ is a minimal vertex separator of $G$ if and only if there exists $Q_i Q_j \in E(T)$ such that $S = Q_i \cap Q_j$.

**Exercise 2.7** For a subtree graph $G$ with clique tree $T$, show that the multiplicity of each $Q_i \cap Q_j$ in the multiset $\{Q_i \cap Q_j : Q_i Q_j \in E(T)\}$ equals one fewer than the number of components in the subgraph of $G$ induced by those vertices that are adjacent to every vertex in $Q_i \cap Q_j$.

**Exercise 2.8** Construct several clique trees for the chordal graph in Figure 2.2 and then use them to illustrate Exercise 2.7.

**Exercise 2.9** For any graph $G$, define $S \subseteq V(G)$ to be a *minimal vertex weak separator* of $G$ if there exist two vertices in a common component of the subgraph of $G$ induced by $V(G) \setminus S$ such that the distance between the two vertices is greater in that subgraph than in $G$. Call an edge $Q_i Q_j$ of $K^\omega(G)$ a "dominated chord" of a clique tree $T$ of $G$ if $Q_i Q_j \notin E(T)$ and $|Q_i \cap Q_j| < |Q \cap Q'|$ for every $QQ' \in E(T(Q_i, Q_j))$.

Show that $S$ is a minimal vertex weak separator of $G$ if and only if there exists a dominated chord $Q_i Q_j$ of $T$ such that $S = Q_i \cap Q_j$.

The next theorem is from [Buneman, 1974], [Gavril, 1974a], and [Walter, 1978]. Our argument follows [Shibata, 1988].

**Theorem 2.4 (Buneman, Gavril, and Walter)** *A graph is a subtree graph if and only if it is a chordal graph.*

**Proof.** First, suppose $G$ is a subtree graph with clique tree $T$. Arguing toward a contradiction, suppose that $G$ contains an *induced* cycle $C$ whose vertices are, in order, $v_1, \ldots, v_k, v_1$ where $k \geq 4$. Putting $v_0 = v_k$ and $v_{k+1} = v_1$, we know that, for $i \in \{1, \ldots, k\}$, $T_{v_i} \cap T_{v_{i-1}} \neq \emptyset \neq T_{v_i} \cap T_{v_{i+1}}$, but, since $C$ is induced, $T_{v_i} \cap T_{v_j} = \emptyset$ for all other vertices $v_j$ of $C$. Thus there exists a path $\Pi$ through $T$ connecting some vertex of $T_{v_1}$ with some vertex of $T_{v_k}$ and containing along the way vertices from each $T_{v_j}$ with $1 < j < k$. But $v_1$ is also adjacent to $v_k$, so $T_{v_1} \cap T_{v_k} \neq \emptyset$ with $T_{v_1} \cap T_{v_k} \cap Q = \emptyset$ for every vertex $Q$ of $\Pi$ in $T_{v_i}$ where $1 < i < k$. This contradicts $T$ being a tree.

Conversely, suppose $G$ contains no induced cycle larger than a triangle and that $T$ is any maximum spanning tree of $K^w(G)$. Arguing toward a contradiction, suppose that there are nonadjacent vertices $Q$ and $Q'$ of $T$ such that (i) there is some vertex in $T(Q, Q')$ that does not contain $Q \cap Q'$ and, among all such, that (ii) $|Q \cap Q'| = k$ is as large as possible. Moreover, among all such $Q, Q'$, suppose that (iii) $T(Q, Q')$ is as short a path as possible. Say $T(Q, Q')$ is $Q = Q_1, Q_2, \ldots, Q_{p-1}, Q_p = Q'$, where $p \geq 3$.

For each $i \in \{1, \ldots, p-1\}$, define $R_i = (Q_i \cap Q_{i+1}) \setminus (Q \cap Q') \subseteq V(G)$. Let each $|Q_i \cap Q_{i+1}| = k_i$ ($1 \leq i < p$). Since $T$ is a maximum spanning tree for $K^w(G)$ and $QQ' \notin E(T)$, each $k_i \geq k$. By (iii), $Q \cap Q' \not\subseteq Q_i$ for each $i \in \{2, \ldots, p-1\}$. Therefore, each $R_i \neq \emptyset$. Since $R_i \cap R_{i+1} \subseteq Q_{i+1}$, the subgraph of $G$ induced by $\cup R_i$ is connected and we can pick a shortest path $x_1, x_2, \ldots, x_q$ therein such that $x_1 \in R_1$ and $x_q \in R_{p-1}$. For each $v \in Q \cap Q'$, $v$ will be adjacent to $x_1$ and $x_q$ and so $v, x_1, \ldots, x_q, v$ will be a cycle in $G$. Since the path $x_1, x_2, \ldots, x_q$ was chosen to be shortest and since $G$ has no induced cycles larger than triangles, each $x_i$ must be adjacent to $v$. Since $v$

was chosen arbitrarily from $Q \cap Q'$, it must be that, for each $i \in \{1, \ldots, q-1\}$, there is a maxclique $S_i$ of $G$ containing $\{x_i, x_{i+1}\} \cup (Q \cap Q')$. Set $S_0 = Q$ and $S_q = Q'$, and note that each $S_i \cap S_{i+1} \supseteq (Q \cap Q') \cup \{x_{i+1}\}$, so $|S_i \cap S_{i+1}| > k$. So by (ii), $S_i \cap S_{i+1}$ is contained in each vertex along $T(S_i, S_{i+1})$ for each $i \in \{0, \ldots, q-1\}$. Thus $Q \cap Q' \subseteq S_i \cap S_{i+1}$ is contained in each vertex along $T(Q, Q')$, contradicting (i). $\qquad \square$

See [Hsu & Ma, 1991] for a linear-time algorithm for finding a clique tree of a chordal graph. Other authors pay attention to what sorts of clique trees a chordal graph can have. For instance, [Blair & Peyton, 1994] gives a linear-time algorithm for finding minimum diameter clique trees of a chordal graph, while [Lih, 1993] investigates finding clique trees that have paths to which all vertices are close. [Lin, McKee, & West, to appear] investigates clique trees having a minimum number of leaves, and [Prisner, 1992] studies chordal graphs that have clique trees with only three leaves. Chapter 3 is devoted to chordal graphs that have clique trees with only two leaves.

[Chen & Lih, 1990] and [Bandelt & Prisner, 1991] characterize chordal graphs whose clique graph is not chordal and show that if $G$ is chordal then $K(K(G))$ is chordal. Section 7.5 is devoted to the clique graphs of chordal graphs.

**Exercise 2.10 (Chen & Lih and Bandelt & Prisner)** Give an example of a chordal graph of order eight whose clique graph is not chordal.

[Raychaudhuri, 1988] gives a polynomial algorithm for finding the intersection number of a chordal graph.

## 2.2 Other Characterizations

One measure of the richness of chordal graph theory is the large number of different characterizations of chordal graphs in the literature; see Theorem 7.47, [Benzaken, Crama, Duchet, Hammer, & Maffray, 1990], and [Bakonyi & Johnson, 1996] for just a few examples. This section considers several standard characterizations, but because of our focus on clique trees and intersection graphs our proofs are not necessarily the standard ones.

**Exercise 2.11** (see [Dirac, 1961]) Show that a graph is chordal if and only if every minimal vertex separator is complete.

We need two standard definitions for Theorem 2.5, from [Fulkerson & Gross, 1965] and [Rose, 1970]. A vertex is a *simplicial* vertex of a graph if its neighbors induce a complete graph (which, remember, includes the case of the null graph). Equivalently, a vertex is simplicial if it is in a unique maxclique. An ordering $\langle v_1, \ldots, v_n \rangle$ of all the vertices of $G$ is a *perfect elimination ordering* of $G$ if, for each $i \in \{1, \ldots, n\}$, $v_i$ is a simplicial vertex of the subgraph induced by $\{v_i, \ldots, v_n\}$.

**Example 2.5** In the graph on the left in Figure 2.1, vertices 1, 2, 3, and 7 are the simplicial vertices. The vertices have been labeled so that their numerical ordering is one possible perfect elimination ordering.

**Theorem 2.5 (Fulkerson & Gross and Rose)** *A graph is chordal if and only if it has a perfect elimination ordering.*

**Proof.** First, suppose $G$ is a subtree graph with clique tree $T$. We argue by induction on the order of $T$ with the result trivial when the order is one. Suppose $Q$ is any maxclique of $G$ corresponding to a leaf of $T$. Since no maxclique can be contained in any other, $Q$ must contain some $v \in V(G)$ that occurs in only that one maxclique, and so $v$ is simplicial. Let $G^-$ result from $G$ by removing $v$, and let $T^-$ result from $T$ by removing $v$ from each vertex of $T$. Then $G^-$ is still a chordal graph, since it has tree representation $T^-$. By inductive hypothesis, $Q^-$ has a perfect elimination ordering that, when $v$ is inserted at the beginning, makes a perfect elimination ordering for $G$.

Conversely, suppose $\langle v_1, \ldots, v_n \rangle$ is a perfect elimination ordering for $G$. We argue by induction on $n$ with the result trivial when $n = 1$. Suppose $Q$ is the maxclique of $G$ consisting of $v_1$ and all its neighbors. Let $G^-$ be the subgraph of $G$ induced by $\{v_2, \ldots, v_n\}$. Since $\langle v_2, \ldots, v_n \rangle$ is a perfect elimination ordering for $G^-$, the inductive hypothesis implies that there is a clique tree $T^-$ for $G^-$. Notice that $Q^- = Q \backslash \{v_1\}$ will be contained in some vertex $R$ of $T^-$. If $Q^- = R$, then let $T$ result by simply inserting $v_1$ into $R$; if $Q^-$ is properly contained in $R$, then let $T$ result by creating a new vertex $Q$ and making it adjacent to $R$. In either case, $T$ is a tree representation for $G$. □

**Exercise 2.12** Show that a graph is chordal if and only if every induced subgraph has a simplicial vertex.

**Exercise 2.13** Show that finding perfect elimination orderings is "foolproof" in the sense that, if $G$ has a perfect elimination ordering, then taking *any* simplicial vertex $v$ of $G$ as a first vertex, then *any* simplicial vertex of the subgraph induced by $V(G) \backslash \{v\}$ as the second, and so on, will always result in a perfect elimination ordering of $G$.

**Exercise 2.14** Show how a perfect elimination ordering for $G$ can be used to give a direct construction of a clique tree for $G$.

We conclude this section with a characterization from [Tarjan & Yannakakis, 1984] that can be implemented in $\mathcal{O}(|V(G)|+|E(G)|)$ time; see also [Golumbic, 1984] and [Shier, 1984]. A *maximum cardinality search* "marks" the vertices of $G$ as follows: First mark an arbitrary vertex; then repeatedly mark any previously unmarked vertex having as many marked neighbors as possible. Stop when all vertices have been marked.

**Example 2.5 (continued)** In the graph in Figure 2.1, taking the vertices in the *opposite* of their numerical order shows one possible order in which they might be marked by a maximum cardinality search. If vertices 5, 6, and 7 (in any order) are the first three marked, then the remaining vertices must be marked in the order $4, 3, 2, 1$.

**Theorem 2.6 (Tarjan & Yannakakis)** *A graph $G$ is chordal if and only if in some maximum cardinality search of $G$, as each vertex becomes marked, its previously marked neighbors are pairwise adjacent in $G$. Moreover, this is equivalent to, in every maximum cardinality search of $G$, as each vertex becomes marked, its previously marked neighbors are pairwise adjacent in $G$.*

**Proof.** If some maximum cardinality search marks the vertices of $G$ in the order $v_1, \ldots, v_n$ such that the neighbors of $v_i$ among $v_1, \ldots, v_{i-1}$ are pairwise adjacent in $G$, then $\langle v_n, \ldots, v_1 \rangle$ is automatically a perfect elimination ordering the $G$, and so $G$ is chordal by Theorem 2.5.

Conversely, suppose $G$ is connected and chordal with clique tree $T$. Suppose a maximum cardinality search marks the vertices of $G$ in the order $v_1, \ldots, v_n$. (We show how maximum cardinality search locates maxcliques of $G$.) No matter which $v_1$ was chosen, vertices $v_1, \ldots, v_k$ (for some $k \le n$) will form a maxclique $Q$ of $G$, because of always marking a vertex that is adjacent to as many previously marked vertices as possible, and so $\{v_1, \ldots, v_k\} = Q \in V(T)$; for the purpose of this proof, call such a vertex $Q$ a "saturated vertex" of $T$. Since $T$ is a maximum spanning tree of $K^w(G)$ by Theorem 2.3, the next vertex $v$ marked in $G$ will occur in some neighbor $Q'$ of $Q$ in $T$ for which $Q \cap Q'$ (the previously marked vertices that $v$ is adjacent to) is as large as possible. Any unmarked vertices occurring in $Q'$ will now be adjacent to more than $|Q \cap Q'|$ previously marked vertices, and so these will be marked next, making $Q'$ saturated. This process continues to make saturated vertices of $T$ one at a time, with the vertices saturated at any time always forming a subtree of $T$. Since each newly marked vertex of $G$ is always in the same maxclique as its previously marked neighbors, these neighbors will be pairwise adjacent.                                      □

**Exercise 2.15** Suppose $G$ is chordal. The first paragraph of the proof of Theorem 2.6 shows that every maximum cardinality search of $G$ corresponds to a reversed perfect elimination ordering of $G$. Show by example that the converse fails—that a perfect elimination ordering of a chordal graph need not correspond to a reversed maximum cardinality search marking.

**Exercise 2.16 (Blair, England, & Thomason)** Prim's algorithm constructs a(ll) maximum spanning tree(s) of a weighted graph by starting at an arbitrary vertex and repeatedly choosing an edge of largest weight that joins a vertex already in the tree with a vertex not yet in the tree. ([Tarjan, 1983] and [Graham & Hell, 1985] contain detailed analysis of both the Kruskal and Prim algorithms.) Discuss how the second paragraph of the proof of Theorem 2.6 illustrates the central theme of [Blair & Peyton, 1993]: that "the maximum cardinality search algorithm is just Prim's algorithm in disguise."

See [Panda, 1996] for deeper discussion of maximum cardinality-type algorithms, and [Simon, 1995] for the role of minimal vertex separators in maximum cardinality-type search algorithms on chordal graphs. [Galinier, Habib, & Paul, 1995] contains more information on clique trees and their role in algorithms. [Kumar & Veni Madhavan, 1989] presents a simple linear-time algorithm for testing the planarity of a chordal graph based on a chordal graph being planar if and only if it is $K_5$-free and each 3-vertex minimal vertex separator has multiplicity one.

## 2.3    Tree Hypergraphs

Continuing the discussion of section 1.6, a hypergraph $(X, \mathcal{E})$ is a *tree hypergraph* if there is a tree $T$ with $X = V(T)$ such that, for each $S_i \in \mathcal{E}$, there is a subtree $T_i$ of $T$ with $V(T_i) = S_i$.

**Exercise 2.17** Show that the hypergraph $(\{a, b, c, d\}, \mathcal{E})$ with $\mathcal{E} = \{\{a\}, \{c\}, \{b, d\}, \{a, b, d\}, \{a, b, c, d\}\}$ is a tree hypergraph, and that the tree $T$ in the definition can be any tree with vertex set $\{a, b, c, d\}$ so long as it contains the edge $bd$ and one of the edges $ab, ad$.

Clearly, the line graph $\Omega(\mathcal{E})$ of a tree hypergraph $(X, \mathcal{E})$ is a subtree graph, and so is a chordal graph by Theorem 2.4. The next example, however, shows that being a tree hypergraph requires more than just having a chordal line graph.

**Example 2.6** The hypergraph $(\{1, 2, 3\}, \mathcal{E})$ having $\mathcal{E} = \{\{1, 2\}, \{1, 3\}, \{2, 3\}\}$ has a chordal line graph ($\cong K_3$), yet is not a tree hypergraph; the following exercise shows that (at least part of) the problem is that $\mathcal{E}$ does not satisfy the Helly condition.

**Exercise 2.18** Show that every tree hypergraph is a Helly hypergraph.

The following theorem appeared independently in [Duchet, 1978], [Flament, 1978], and [Slater, 1978]; our argument follows Slater's.

**Theorem 2.7 (Duchet, Flament, and Slater)** *A hypergraph is a tree hypergraph if and only if it is a Helly hypergraph with a chordal line graph.*

**Proof.** We have already observed the implication one way. For the converse, suppose $(X, \mathcal{E})$ is a Helly hypergraph and its line graph $G = \Omega(\mathcal{E})$ is chordal. Say $\mathcal{E} = \{S_1, \ldots, S_m\}$. We argue by induction on $m$. For the $m = 1$ basis, $(X, \mathcal{E})$ is a tree hypergraph for which $T$ can be *any* tree with vertex set $S_1$. Suppose $m > 1$. Since $G$ is chordal, Theorem 2.5 allows us to reorder the $S_i$'s as necessary so that $S_1$ is a simplicial vertex of $G$ and $\{S_1, \ldots, S_k\}$ induces the unique maxclique of $G$ that contains $S_1$. We can assume $k \geq 2$ since if $k = 1$, meaning that $S_1$ is an isolated vertex in $G$, then the remainder of the argument becomes trivial. By the Helly condition, there is some $x \in S_1 \cap \cdots \cap S_k$. Put $S_1' = S_1 \setminus \{x\}$ and $S_i' = S_i \setminus S_1'$ when $i \geq 2$. Note that $k < j \leq m$ implies $S_1 \cap S_j = \emptyset$ and $S_j' = S_j$. Suppose $i$ and $j$ are such that $2 \leq i < j \leq m$. If $S_i' \cap S_j' \neq \emptyset$, then $S_i \cap S_j \supseteq S_i' \cap S_j' \neq \emptyset$. If $S_i \cap S_j \neq \emptyset$, then either $j \leq k$ and $x \in S_i' \cap S_j' \neq \emptyset$, or $j > k$ and $S_i' \cap S_j' = (S_i \setminus S_1') \cap S_j = S_i \cap S_j \neq \emptyset$ since $S_1 \cap S_j = \emptyset$. Thus $S_i \cap S_j \neq \emptyset$ if and only if $S_i' \cap S_j' \neq \emptyset$. In this way, $\{S_2', \ldots, S_m'\}$ satisfies the Helly condition and $\Omega(\{S_2', \ldots, S_m'\}) \cong \Omega(\{S_2, \ldots, S_m\})$ is chordal. So by the induction hypothesis, $(X \setminus S_1', \{S_2', \ldots, S_m'\})$ is a tree hypergraph with respect to some tree $T'$. Form $T$ from $T'$ by adding, for each element of $S_1'$, a new vertex of degree one adjacent to $x$. Then each $S_i$ is the vertex set of a subtree of $T$ and $V(T) = X$. $\square$

**Exercise 2.19** Show that a graph is chordal if and only if it is the line graph of a tree hypergraph.

Let $H = (X, \mathcal{E})$ be a hypergraph. A *partial hypergraph* of $H$ is a hypergraph $H' = (X', \mathcal{E}')$ where $\mathcal{E}' \subseteq \mathcal{E}$ and $X' = \cup_{S' \in \mathcal{E}'} S'$. If $A \subseteq X$, the *subhypergraph of $X$ induced by $A$* is the hypergraph $H_A = (A, \mathcal{E}_A)$ where $\mathcal{E}_A = \{S \cap A : S \in \mathcal{E}\}$.

While sections 2.1 and 2.2 show that every induced subgraph of a subtree graph is itself a subtree graph, the following example shows that this hereditary property fails for tree hypergraphs.

**Example 2.7** Consider the tree hypergraph $(\{1, 2, 3, 4\}, \mathcal{E})$ in which $\mathcal{E} = \{\{1, 2, 3\}, \{1, 2, 4\}, \{2, 3, 4\}\}$. If $A = \{1, 3, 4\}$, then Theorem 2.7 shows that $H_A$ is not a tree hypergraph.

The *dual hypergraph* $H^* = (X^*, \mathcal{E}^*)$ of a hypergraph $H = (X, \mathcal{E})$ has $X^* = \mathcal{E}$ with $\mathcal{E}^* = \{S_x^* : x \in X\}$ where each $S_x^* = \{S \in \mathcal{E} : x \in S\}$. Note that $H^{**} \cong H$. Given a graph $G$, the *clique hypergraph* of $G$ is the hypergraph $(V(G), \mathcal{E})$ where $\mathcal{E}$ is the set of all maxcliques of $G$.

**Exercise 2.20** Show that a graph $G$ is chordal if and only if $H^*$ is a tree hypergraph where $H$ is the clique hypergraph of $G$.

**Exercise 2.21** Show that the dual of a subhypergraph of the hypergraph $H$ is isomorphic to a partial hypergraph of $H^*$.

Section 2.4.2 will sketch an application of tree hypergraphs in database theory. See [Naiman & Wynn, 1992] for an application in probability theory of duals of tree hypergraphs (called "generalized simple tubes" there).

A *cycle of length $k$* in the hypergraph $H = (X, \mathcal{E})$ is a sequence $v_1, S_1, v_2, S_2, \ldots, S_k, v_1$ where $S_1, \ldots, S_k$ are distinct edges, $v_1, \ldots, v_k$ are distinct vertices, $v_i, v_{i+1} \in S_i$ for all $i = 1, \ldots, k - 1$, and $v_k, v_1 \in S_k$. A *totally balanced hypergraph* is a hypergraph in which every cycle of length greater than two contains an edge $S_i$ that contains at least three of the vertices $v_1, \ldots, v_k$ of the cycle.

**Exercise 2.22** Suppose $H$ is any totally balanced hypergraph. Show that $H^*$ and all the partial hypergraphs and subhypergraphs of $H$ are also totally balanced and that $H$ must be a Helly hypergraph.

The following theorem can be found in [Lehel, 1983, 1985] and [Ryser, 1969].

**Theorem 2.8** *A hypergraph is totally balanced if and only if each of its subhypergraphs is a tree hypergraph.*

**Proof.** Suppose $H$ is a totally balanced hypergraph. Exercise 2.22 shows that every subhypergraph of $H$ is a totally balanced Helly hypergraph that, by the definition of totally balanced, has a chordal line graph. Theorem 2.7 then implies that every subhypergraph of $H$ is a tree hypergraph.

Conversely, suppose every subhypergraph of $H$ is a tree hypergraph, yet suppose $H$ has a cycle of length three with none of its edges containing three vertices of the cycle. If this cycle has length three, then those three vertices would induce a subhypergraph of $H$ that is not a Helly hypergraph; if it has length greater than three, then its vertices would induce a subhypergraph of $H$ whose line graph is not chordal. Either case contradicts Theorem 2.7. $\square$

A hypergraph is a *strong Helly hypergraph* if each of its subhypergraphs is a Helly hypergraph. Compare the following with the Gilmore criterion in Exercise 1.23.

**Theorem 2.9 (Lehel)** *A hypergraph $H = (X, \mathcal{E})$ is a strong Helly hypergraph if and only if, for all $u, v, w \in X$, there exists $x \in \{u, v, w\}$ such that every edge in $\mathcal{E}$ that contains at least two of $u, v, w$ also contains $x$.*

**Proof.** This follows from applying Exercise 1.23 to all the subhypergraphs induced by distinct $u, v, w \in X$. $\square$

**Corollary 2.10 (Lehel)** *If a hypergraph is totally balanced, then it is both a tree hypergraph and a strong Helly hypergraph.*

**Proof.** Suppose $H$ is totally balanced. Theorem 2.8 implies $H$ is a tree hypergraph. Since every cycle of $H$ of length three has an edge containing at least three vertices of the cycle, Theorem 2.9 can be used to show that $H$ is strong Helly. $\square$

**Exercise 2.23** Use the hypergraph $H = (\{0, 1, 2, 3, 4\}, \{S_1, S_2, S_3, S_4\})$ with $S_1 = \{0, 2, 3\}$, $S_2 = \{0, 3, 4\}$, $S_3 = \{0, 1, 4\}$, and $S_4 = \{0, 1, 2\}$ to show that the converse to Corollary 2.10 fails.

Totally balanced hypergraphs also play an important role with respect to "strongly chordal graphs," as discussed in section 7.12, as do strong Helly hypergraphs with respect to "hereditary clique-Helly graphs" in section 7.5.

## 2.4   Some Applications of Chordal Graphs

Each of the following subsections is merely a brief sketch of one application of chordal graphs. As an example of a type of application that is not
represented in our collection, [Chandrasekaran & Tamir, 1982] studies the
location of "supply centers" on a network having a tree structure. Section 3.4 consists of additional examples when the tree representations are
required to be paths.

### 2.4.1   Applications to Biology

Let $S$ denote a given set of molecular sequences, where each sequence corresponds to a *taxon* (an organism). For simplification, assume each sequence
in $S$ has length $k$ and is built from the four letter alphabet $B = \{A,C,G,T\}$;
thus each corresponds to a DNA sequence on the four *bases* A, C, G, and
T. (Protein sequences are similarly built from a 20 letter alphabet.) In the
language of numerical taxonomy, the taxa (organisms) are described by $k$
*characters*, each having one of four possible *states*. These characters can be
represented as functions $f_1, \ldots, f_k$ where $f_i : S \to B$ with $f_i(x)$ the base at
position $i$ for taxon $x \in S$. Note that each character $f_i$ induces a partition
of the set $S$ of taxa into at most four nonempty equivalence classes, the
preimages of the bases in $B$.

   *Compatibility analysis* seeks to find collections of characters from among
$f_1, \ldots, f_k$ that are *compatible* (consistent) in that there exists a tree $T$ with
$S \subseteq V(T)$ on which, for each $f_i$ in the collection, each equivalence class
of $f_i$ corresponds to a subtree of $T$. If a collection of characters is not
compatible—if there is no such tree—then insofar as the evolutionary history
for $S$ is a tree, the true evolutionary history for $S$ is not reflected in those
characters. Thus compatibility analysis, considered as a consistency test, is
a valuable method in that it can tell us something definite (albeit negative)
about evolutionary aspects of certain characters.

   Now suppose $B$ is any finite set of states and each character $f_i$ corre-
sponds to a partition $P_i = \{X_1^{(i)}, \ldots, X_{m_i}^{(i)}\}$ of $S$, with each $m_i \leq |B|$ and
$X_s^{(i)} \neq \emptyset$. Note that it is possible to have $X_s^{(i)}$ and $X_t^{(j)}$ equal as sets, even
though $i \neq j$. Define the *partition intersection graph* $\Omega(P_1, \ldots, P_k)$, $k \geq 2$,
to have vertices $\{X_1^{(1)}, \ldots, X_{m_1}^{(1)}, \ldots, X_1^{(k)}, \ldots, X_{m_k}^{(k)}\}$, with $X_s^{(i)}$ adjacent to
$X_t^{(j)}$ if and only if $i \neq j$ and $X_s^{(i)} \cap X_t^{(j)} \neq \emptyset$. Notice that $\Omega(P_1, \ldots, P_k)$ is a
$k$-partite graph with chromatic number $k$, and each taxon $x \in S$ corresponds
to a maxclique of order $k$. [McMorris & Meacham, 1983] characterizes all
graphs that arise this way.

**Theorem 2.11 (McMorris & Meacham)** *A graph with chromatic number $k > 1$ is a partition intersection graph if and only if it has no isolated vertices and has an edge clique cover, each member of which has order $k$.*

**Proof.** Assume $G$ has chromatic number $k > 1$.

First suppose $G \cong \Omega(P_1, \ldots, P_k)$. Check that (1) each $X_s^{(i)} \in V(G)$ has at least $k - 1 \geq 1$ neighbors; (2) for each taxon $x \in S$ and each $\mathsf{B} \in B$, $\{X_s^{(i)} : f_i(x) = \mathsf{B}\}$ induces a complete subgraph of order $k$ in $G$; and (3) the family of all such induced subgraphs is an edge clique cover of $G$.

Conversely, suppose $\mathcal{E} = \{Q_1, \ldots, Q_m\}$ is an edge clique cover for $G$ where each $|Q_j| = k$, $G$ has been *properly $k$-colored* (meaning that no two adjacent vertices have the same color), and no vertex of $G$ is isolated. For each $v \in V(G)$, set $\mathcal{E}_v = \{Q_j : v \in Q_j\}$. Since each $v \in V(G)$ is on at least one edge of $G$ and that edge is in at least one member of the edge clique cover $\mathcal{E}$, each $\mathcal{E}_v \neq \emptyset$. Suppose $\{v_1, \ldots, v_\ell\}$ is any one of the $k$ color classes. Then each $Q_i \in \mathcal{E}$ will contain exactly one of $v_1, \ldots, v_\ell$ and so will be contained in exactly one of $\mathcal{E}_{v_1}, \ldots, \mathcal{E}_{v_\ell}$. Since $\mathcal{E}_{v_1}, \ldots, \mathcal{E}_{v_\ell}$ are disjoint subsets of $\mathcal{E}$, they partition $\mathcal{E}$. So each of the $k$ color classes corresponds to one of $k$ partitions $P_1, \ldots, P_k$ of $\mathcal{E}$. Moreover, $uv \in E(G)$ if and only if, for some $i$, both $u, v \in Q_i$ where $u$ and $v$ are in different color classes. But this is equivalent to $Q_i \in \mathcal{E}_u \cap \mathcal{E}_v$ with $\mathcal{E}_u, \mathcal{E}_v$ in different partitions among $P_1, \ldots, P_k$, which in turn is equivalent to $\mathcal{E}_u \mathcal{E}_v$ being an edge of $\Omega(\{P_1, \ldots, P_k\})$. Hence $G \cong \Omega(\{P_1, \ldots, P_k\})$. $\square$

In the case of just two characters $f_i$ and $f_j$, being compatible is equivalent to the bipartite graph $\Omega(P_i, P_j)$ being acyclic. Pairwise compatibility can be used to construct a *compatibility graph*, using the characters as vertices with adjacency corresponding to pairwise compatibility. Compatibility analysis seeks the largest collections of compatible characters. In the special case where every character has only two possible states, [McMorris, 1977] shows that maxcliques of the compatibility graph correspond to maximal compatible collections of characters. See also [Gusfield, 1991]. However, this fails in general, as shown in [Fitch, 1977] and by an infinite family of examples in [Meacham, 1983].

By assigning each character $f_i$ (and so each corresponding partition $P_i$ of $S$) a color $i$ and coloring each vertex $X_s^{(i)}$ of $\Omega(P_1, \ldots, P_k)$ with color $i$, we have a *chromatic chordal completion problem*, as in [Buneman, 1974]: Given a graph whose vertices are properly $k$-colored, determine whether edges can be added between vertices of different colors in order to make the graph

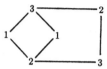

Figure 2.3: *A graph with no chromatic chordal completion.*

chordal. Edges can be added in this way if and only if the collection of characters is compatible.

**Example 2.8** The graph shown in Figure 2.3, 3-colored with "colors" 1, 2, and 3, cannot be made chordal by adding edges between vertices of different colors: The second 2-3 edge would have to be added to eliminate the length-four "1,2,1,3" cycle, creating a new length-four "2,3,2,3" cycle that could not be eliminated.

Finding the complexity of the chromatic chordal completion problem was posed in [McMorris & Meacham, 1983]. Note that the chromatic restriction is important since an arbitrary uncolored graph can obviously be made into a chordal graph; thus the only problem in the single color case is to find minimal and minimum such sets of edges. See [Rose & Tarjan, 1975] and [Rose, Tarjan, & Lueker, 1976] for the complexity of these problems.

Recently, there has been a lot of activity in assessing the computational complexity of all the variations on the chromatic chordal completion problem. See [Bodlaender & Kloks, 1993], [McMorris, Warnow, & Wimer, 1994], [Agarwala & Fernández-Baca, to appear], [Kannan & Warnow, 1992, 1994], [Indury & Schaeffer, 1993], and [Bodlaender, Fellows, & Warnow, 1992].

There is still a lot of theoretical work to do before these results can be useful for compatibility analysis. For example, how can a largest set of compatible characters be selected when chromatic chordal completion is impossible?

## 2.4.2   Applications to Computing

The application discussed in subsection 2.4.1 is an example of a general "filiation" problem, determining whether certain data or objects are compatible with arrangement in a tree pattern. Corresponding "seriation" problems, with arrangement in a linear pattern, will be discussed in Chapter 3.

Another filiation application occurs in computer science. A *database scheme* can be thought of as a collection of tables—for instance, a manager's table with columns for "employee name," "social security number," "position," "job skills," etc.; a payroll table with columns for "social security number," "date of employment," "salary," etc.; a receptionist's table with columns for "employee name," "telephone extension," "hours," etc.; and so on—with the tables called *relations* and their columns called *attributes*.

Suppose a database scheme consists of a family $\mathcal{R}$ of relations and a set $X$ of attributes (so $(X, \mathcal{R})$ is a hypergraph). This is an *acyclic database scheme* if the relations in $\mathcal{R}$ can be arranged as the vertices of a tree, commonly called a *join tree*, such that the vertices containing any given attribute induce a subtree. Join trees are like tree representations, and paths within the join tree constitute unique retrieval paths for data. Having a join tree representation is one of a large number of desirable properties of database schemes— matters of consistency, efficiency, and compatibility—that are shown to be equivalent to each other in [Beeri, Fagin, Maier, & Yannakakis, 1983], which also cites evidence that database schemes that possess these desirable properties are "sufficiently general to encompass most 'real-world' situations." [Golumbic, 1988] provides a simple introduction.

In terms of graphs, define $G = G(\mathcal{R})$ to have $V(G) = X$ with $E(G) = \{xy : x, y \in R \in \mathcal{R}\}$ (so $\mathcal{R}$ is an edge clique cover of $G(\mathcal{R})$).

**Proposition 2.12** *A database scheme $\mathcal{R}$ is an acyclic database scheme if and only if each complete subgraph of $G(\mathcal{R})$ is contained in a common member of $\mathcal{R}$ and $G(\mathcal{R})$ is a chordal graph.* □

In terms of hypergraphs, $\mathcal{R}$ is an acyclic database scheme if and only if the dual of the hypergraph $(X, \mathcal{R})$ is a tree hypergraph; Proposition 2.12 then corresponds to Theorem 2.7. (Warning: There are many different notions of "cycle" and "acyclic" in use for hypergraphs, and being "acyclic" very often means something different from not having a "cycle"; acyclic database schemes correspond to what are often called "$\alpha$-acyclic" hypergraphs.)

Subsection 2.4.4 will mention a somewhat related role of chordal graphs connected with expert systems.

Here is another, completely different application in computing: Many problems that are NP-complete in general become tractable, sometimes even solvable in linear time, when restricted to chordal graphs. While for many people this is the most important application of chordal graphs, it often

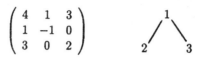

Figure 2.4: *A matrix M and its graph G(M).*

involves little of the specific nature of chordal graphs as intersection graphs—other highly structured families can do as well. [Klein, 1996] discusses similar computational concerns for parallel computing.

[Chung & Mumford, 1994] is an example of such computational concerns, dealing with problems arising in computer vision. The specific problem faced is to determine bounds on the number of edges needed to be added to make a nonchordal graph chordal—the *minimum fill-in problem*—and so susceptible to more efficient algorithms. [Kloks, Bodlaender, Müller, & Kratsch, 1993], [Kloks & Kratsch, 1994], and [Parra & Scheffler, 1995, 1997], for instance, discuss how to use the minimal vertex separators of the nonchordal graph to determine how to add a minimal set of edges. Similar concerns also arise with sparse matrix computation in subsection 2.4.3 and maximum likelihood estimation in subsection 2.4.4.

### 2.4.3  Applications to Matrices

Gaussian elimination on an $n \times n$ matrix $M = (m_{ij})$ involves the choice of a nonzero *pivot* entry $m_{ij}$, then using elementary row and column operations to change $m_{ij}$ into 1 and all other $i$th row and $j$th column entries into 0. An *elimination scheme* is a sequence of $n$ pivots used to reduce a matrix to the identity matrix, and a *perfect elimination scheme* has the further property that no zero entry is ever made nonzero along the way. Perfect elimination schemes minimize both computation and data storage.

The *graph of M*, denoted $G(M)$, has vertex set $\{1, \ldots, n\}$, with vertices $i \neq j$ adjacent if and only if either $m_{ij} \neq 0$ or $m_{ji} \neq 0$.

**Example 2.9** Figure 2.4 shows a matrix and its graph.

Proposition 2.13 is from [Rose, 1970] and has led to much further work; see [Rose, 1972], Chapter 12 of [Golumbic, 1980], [Golumbic, 1984], and various papers in [George, Gilbert, & Liu, 1993].

**Proposition 2.13 (Rose)** *A symmetric matrix $M$ with nonzero diagonal entries has a perfect elimination scheme using the diagonal entries as pivots if and only if $G(M)$ is chordal.*

**Proof sketch.** Pivoting on $m_{ii}$ results in removing all the edges incident to vertex $i$ in $G(M)$ and simultaneously creating a new edge $hj$ whenever $m_{hi} \neq 0 \neq m_{ij}$ but $m_{hj} = 0$, as below:

$$\begin{pmatrix} & \vdots & & \vdots & \\ \dots & m_{ii} & \dots & m_{ij} & \dots \\ & \vdots & & \vdots & \\ \dots & m_{hi} & \dots & 0 & \dots \\ & \vdots & & \vdots & \end{pmatrix} .$$

(Other entries might also inadvertently become zero in M, and so other edges disappear from $G(M)$.) Hence no zero entry is made nonzero in $M$ precisely when every two neighbors of $i$ ($h$ and $j$ above) are adjacent in $G(M)$; equivalently, when $i$ is a simplicial vertex. (For instance, in the matrix in Example 2.9, you could pivot on either of the entries $-1$ or $2$ but not on $4$.) A perfect elimination scheme on the diagonal entries of $M$ thus corresponds to a perfect elimination ordering for $G(M)$, and such a perfect elimination scheme exists if and only if $G$ is chordal. □

The remainder of this subsection will describe a less practical, but more surprising, appearance of chordal graphs in matrix analysis.

It is trivial to compute the determinant of $A$ when $A^{-1}$ is a diagonal matrix: $\det A = \prod_{i=1}^{n} a_{ij}$. There are also simple methods to compute $\det A$ whenever $A^{-1}$ is known to be "tridiagonal," meaning that the entries $b_{ij}$ of $A^{-1}$ are zero whenever $|i - j| > 1$. Observe that $G(A^{-1})$ is then a union of paths. In the early 1980s, this was generalized to larger families of matrices, including, in [Klein, 1982], when $A^{-1}$ is "treediagonal" (meaning that $G(A^{-1})$ is a tree). This work culminated in Proposition 2.14 from [Barrett & Johnson, 1984] ("reinventing" chordal graphs). See [Barrett, Johnson, & Lundquist, 1989] and [Johnson, 1990] for more recent surveys of where this led next. For any $n \times n$ matrix $M$ and any set $S \subseteq \{1, \dots, n\}$ (or any subgraph $S$ of $G(M)$), let $M[S]$ denote the submatrix determined by those rows and columns of $M$ indexed by the elements of $S$.

$$\begin{pmatrix} ? & ? & ? & 0 & 0 \\ ? & ? & ? & 0 & ? \\ ? & ? & ? & ? & ? \\ 0 & 0 & ? & ? & 0 \\ 0 & ? & ? & 0 & ? \end{pmatrix}$$

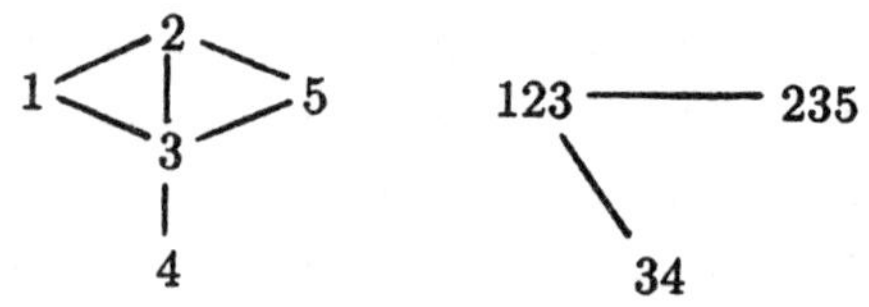

Figure 2.5: *A family of matrices having the same chordal graph $G$, with one clique tree for $G$.*

**Proposition 2.14 (Barrett & Johnson)** *If $G(A^{-1})$ is chordal with clique tree $T$, then*

$$\det A = \frac{\prod_{Q \in V(T)} \det A[Q]}{\prod_{Q_i Q_j \in E(T)} \det A[Q_i \cap Q_j]}, \tag{2.2}$$

*provided the denominator is nonzero.*

**Example 2.10** Suppose $A$ is any matrix whose inverse $A^{-1}$ is as shown in Figure 2.5, with the ? entries unspecified (possibly zero). The graph $G(A^{-1})$ is shown in the middle, and one of the two possible clique trees $T$ for $G(A^{-1})$ is shown at the right; for either clique tree, $E(T) = \{\{2,3\}, \{3\}\}$. Formula (2.2) becomes

$$\det A = \frac{\det A[\{1,2,3\}] \cdot \det A[\{2,3,5\}] \cdot \det A[\{3,4\}]}{\det A[\{2,3\}] \cdot \det A[\{3\}]}$$

$$= \frac{\det \begin{pmatrix} a_{11} & a_{12} & a_{13} \\ a_{21} & a_{22} & a_{23} \\ a_{31} & a_{32} & a_{33} \end{pmatrix} \cdot \det \begin{pmatrix} a_{22} & a_{23} & a_{25} \\ a_{32} & a_{33} & a_{35} \\ a_{52} & a_{53} & a_{55} \end{pmatrix} \cdot \det \begin{pmatrix} a_{33} & a_{34} \\ a_{43} & a_{44} \end{pmatrix}}{\det \begin{pmatrix} a_{22} & a_{23} \\ a_{32} & a_{33} \end{pmatrix} \cdot a_{34}}.$$

**Proof sketch.** Suppose $G = G(A^{-1})$ is chordal with clique tree $T$. Suppose $L$ is any leaf vertex of $T$ and $R$ is the set of all entries that are in vertices of $T$ other than $L$. Thus $L \cap R$ corresponds to the edge joining $L$ to the rest of $T$. We show that

$$\det A = \frac{\det A[L] \cdot \det A[R]}{\det A[L \cap R]}, \tag{2.3}$$

from which Proposition 2.14 follows inductively.

Suppose, for convenience, that the elements of $L \setminus R$ come first in the matrices $A$ and $A^{-1}$, with those in $L \cap R$ coming next and those not in $L$ coming last. Thus the matrices consist of nonempty blocks as shown below, with two blocks of $A^{-1}$ consisting entirely of zero entries, reflecting that no vertex in $L \setminus R$ is adjacent to any vertex not in $L$:

$$A = \left( \begin{array}{c|c|c} A_{11} & A_{12} & A_{13} \\ \hline A_{21} & A_{22} & A_{23} \\ \hline A_{31} & A_{32} & A_{33} \end{array} \right), \quad A^{-1} = \left( \begin{array}{c|c|c} B_{11} & B_{12} & 0 \\ \hline B_{21} & B_{22} & B_{23} \\ \hline 0 & B_{32} & B_{33} \end{array} \right).$$

It is easy to verify that

$$\det B_{11} \cdot \det B_{33} = \det \left( \begin{array}{c|c} B_{11} & 0 \\ \hline 0 & B_{33} \end{array} \right). \tag{2.4}$$

By a result of Jacobi from 1834 relating minors of $A$ and $A^{-1}$,

$$\det B_{11} \cdot \det A = \det \left( \begin{array}{c|c} A_{22} & A_{23} \\ \hline A_{32} & A_{33} \end{array} \right),$$

$$\det B_{33} \cdot \det A = \det \left( \begin{array}{c|c} A_{11} & A_{12} \\ \hline A_{21} & A_{22} \end{array} \right),$$

and $\det \left( \begin{array}{c|c} B_{11} & 0 \\ \hline 0 & B_{33} \end{array} \right) \cdot \det A = \det A_{22}.$

Multiplying both sides of (2.4) by $(\det A)^2$ and then using these three equalities gives

$$\det \left( \begin{array}{c|c} A_{22} & A_{23} \\ \hline A_{32} & A_{33} \end{array} \right) \cdot \det \left( \begin{array}{c|c} A_{11} & A_{12} \\ \hline A_{21} & A_{22} \end{array} \right) = \det A_{22} \cdot \det A,$$

from which (2.3) follows by dividing through by $\det A_{22}$.                                    □

Perfect Gaussian elimination and determinantal formulas, the two topics of this subsection, can be interrelated as in [Bakonyi, 1992].

**Exercise 2.24** (see [Grone, Johnson, Sá, & Wolkowicz, 1984]) Show that, in any chordal graph, new edges can be inserted one at a time, always maintaining a chordal graph, all the way up to a complete graph.

This exercise is important in another broad topic—"matrix completion problems." See [Johnson, 1990], [Bakonyi & Johnson, 1995], and [Johnson, Jones, & Kroschel, 1995] for examples. [Bakonyi & Johnson, 1996] deals directly with several algebraic characterizations of chordal graphs.

### 2.4.4   Applications to Statistics

The application of chordal graphs to statistics dates from the seminal paper [Darroch, Lauritzen, & Speed, 1980] (another "reinvention" of chordal graphs). Recent textbook introductions to this use of graphs include [Santner & Duffy, 1989], [Wickens, 1989], [Christensen, 1990], [Whittaker, 1990], and [Lauritzen, 1996]. [Khamis & McKee, 1997] is a guide to this literature written for graph theorists, while [McKee & Khamis, 1996] and [Khamis, 1996] present a multigraph approach to some of the same topics. [Lauritzen & Spiegelhalter, 1988], [Pearl, 1988], and [Neapolitan, 1990] move on into the propagation of probabilistic evidence in expert systems.

For a set $\{1, \ldots, d\}$ of *factors*, a *level* $\ell_i$ of factor $i$ is an allowable value of factor $i$. We denote the common occurrence of levels $\ell_1, \ldots, \ell_d$ by the conjunction $\bigwedge\{\ell_i : 1 \le i \le d\}$. A ("$d$-dimensional hierarchical loglinear") *model $M$* consists of a set of *generators*: incomparable subsets of $\{1, \ldots, d\}$. Generators correspond to inclusion-maximal sets of factors having interrelationships taken to be significant within the model. The *interaction graph of $M$*, denoted $G(M)$, has the factors as vertices, with two adjacent whenever the factors are in a common generator; thus the generators of $M$ form an edge clique cover of $G(M)$. If the generators of $M$ are precisely the maxcliques of $G(M)$, then $M$ is called a *graphical model*.

**Example 2.11**  Suppose $d = 7$ where factor 1 corresponds to "sex" with $\ell_1 \in \{$male, female$\}$, factor 2 is "age" with $\ell_2 \in \{$under 30, 30–45, 46–60, over 60$\}$, and so on. Suppose the generators are $\{1, 5\}$, $\{2, 4, 5\}$, $\{3, 4, 5, 6\}$, and $\{4, 5, 6, 7\}$, so $G(M)$ is the graph shown on the left in Figure 2.1. In this model, the interaction of factor 1 ("sex") is not taken as important with factor 2 ("age"), but only with factor 5 (perhaps "occupation").

Choosing which model to apply to observed data involves various statistical techniques that are not of concern here. But not all models are equally easy to make inferences from. Those with particularly desirable properties, usually called *decomposable models*, were shown in [Darroch, Lauritzen, & Speed, 1980] to be precisely those that have chordal interaction graphs.

Suppose a large population is sampled and, for that sample, the number of individuals having each possible combination of levels of factors is determined. The principal advantage of a decomposable model with $g$ generators for these data is that the predictive value of all the data is contained in knowing these numbers for a small number of special sets of factors: the $g$ generators and a certain $g - 1$ intersections of pairs of generators. Proposition 2.15, from [Darroch, Lauritzen, & Speed, 1980], shows how this is done. In this proposition, for any $S \subseteq \{1, \ldots, d\}$, $\hat{p}(\bigwedge\{\ell_i : i \in S\})$ denotes the number of individuals in the sample who have the combination $\bigwedge\{\ell_i : i \in S\}$ of levels of factors in $S$, divided by the total number of individuals in the sample. This proposition actually characterizes decomposable models, and so chordal interaction graphs, and replaces iterative methods that are needed in the general case.

**Proposition 2.15 (Darroch, Lauritzen, & Speed)** *If $G(M)$ is chordal with clique tree $T$, then for every choice $\ell_1, \ldots, \ell_d$ of levels of the factors,*

$$\hat{p}\left(\bigwedge\{\ell_i : i \in V(G)\}\right) = \frac{\prod_{Q \in V(T)} \hat{p}(\bigwedge\{\ell_i : i \in Q\})}{\prod_{Q_i Q_j \in E(T)} \hat{p}(\bigwedge\{\ell_i : i \in Q_i \cap Q_j\})}, \qquad (2.5)$$

*provided the denominator is nonzero.*

**Proof sketch.** Suppose $G = G(M)$ is chordal with clique tree $T$. Suppose $L$ is any leaf vertex of $T$ and $R$ is the set of all factors that are in vertices of $T$ other than $L$. Thus $L \cap R$ corresponds to the edge joining $L$ to the rest of $T$. We show that

$$\hat{p}\left(\bigwedge\{\ell_i : i \in V(G)\}\right) = \frac{\hat{p}(\bigwedge\{\ell_i : i \in L\}) \cdot \hat{p}(\bigwedge\{\ell_i : i \in R\})}{\hat{p}(\bigwedge\{\ell_i : i \in L \cap R\})}, \qquad (2.6)$$

from which Proposition 2.15 follows inductively.

For any set $S$ of factors, let $\bigwedge S$ abbreviate the compound event of each factor $i \in S$ having level $\ell_i$. Then $\hat{p}(\bigwedge\{\ell_i : i \in S\})$ approximates, for the entire population sampled, the joint probability of the compound event $\bigwedge S$; abbreviate this probability by $\Pr[\bigwedge S]$. Then (2.6) corresponds to

$$\Pr\left[\bigwedge V(G)\right] = \frac{\Pr[\bigwedge L] \cdot \Pr[\bigwedge R]}{\Pr[\bigwedge(L \cap R)]}. \qquad (2.7)$$

Since $L \cap R$ corresponds to the edge of $T$ joining $L$ to the rest of $T$, every path from a vertex of $L \backslash R$ to a vertex of $R \backslash L$ passes through a vertex of $L \cap R$. This means that the compound events $\bigwedge(L \backslash R)$ and $\bigwedge(R \backslash L)$ are conditionally independent, conditioning on $\bigwedge(L \cap R)$; in symbols,

$$\Pr\left[\bigwedge(L \backslash R) \,|\, \bigwedge(L \cap R)\right] \cdot \Pr[\bigwedge(R \backslash L) \,|\, \bigwedge(L \cap R)] \qquad (2.8)$$

$$= \Pr[\bigwedge(L \backslash R \cup R \backslash L) \,|\, \bigwedge(L \cap R)].$$

By the definition of conditional probability,

$$\Pr\left[\bigwedge(L \backslash R) \,|\, \bigwedge(L \cap R)\right] = \frac{\Pr[\bigwedge L]}{\Pr[\bigwedge(L \cap R)]},$$

$$\Pr\left[\bigwedge(R\backslash L)\mid \bigwedge(L\cap R)\right] = \frac{\Pr[\bigwedge R]}{\Pr[\bigwedge(L\cap R)]},$$

and

$$\Pr[\bigwedge(L\backslash R \cup R\backslash L)\mid \bigwedge(L\cap R)]$$
$$= \frac{\Pr[\bigwedge(L\backslash R \cup R\backslash L \cup (L\cap R))]}{\Pr[\bigwedge(L\cap R)]} = \frac{\Pr[\bigwedge V(G)]}{\Pr[\bigwedge(L\cap R)]}.$$

Using these three equalities in (2.8) and then multiplying through by $\Pr[\bigwedge(L\cap R)]^2$ gives

$$\Pr\left[\bigwedge L\right]\cdot \Pr\left[\bigwedge R\right] = \Pr\left[\bigwedge V(G)\right]\cdot \Pr\left[\bigwedge(L\cap R)\right],$$

from which (2.7) follows by dividing through by $\Pr[\bigwedge(L\cap R)]$ and using $\Pr[\bigwedge S] \approx \hat{p}(\bigwedge\{\ell_i : i \in S\})$ again.                                      □

Propositions 2.14 and 2.15 are obviously similar in form, each using a product over $V(T)$ divided by a product over $E(T)$. This similarity is no coincidence, as may be sensed from the proof sketches, but a more abstract viewpoint is needed in order to be precise; [Speed & Kiiveri, 1986] and [McKee, 1993] present such viewpoints.

## 2.5   Split Graphs

A graph $G$ is a *split graph* if $V(G)$ can be partitioned into $Q \cup I$, where $Q$ induces a complete graph and $I$ induces an edgeless graph; thus $G$ has $|Q|(|Q| - 1)/2$ edges within $Q$ and anywhere from zero to $|Q| \cdot |I|$ other edges, between $Q$ and $I$. Split graphs were introduced in [Földes & Hammer, 1977a]; also see Chapter 6 of [Golumbic, 1980]. (Warning: [Földes & Hammer, 1977b] gives a different meaning to "split.") They were independently studied in [Tyškevič & Černjak, 1978a, 1978b, 1979]. While split graphs may seem too special to be of interest, the theorem and corollary in this section guarantee the place of split graphs in intersection graph theory.

**Exercise 2.25** Show that the definition of split graphs could have equivalently required that $Q$ be a maxclique of $G$.

**Theorem 2.16 (Földes & Hammer)** *A graph $G$ is a split graph if and only if both $G$ and its complement $\overline{G}$ are chordal.*

**Proof.** First suppose $G$ is a split graph. Then it is easy to see that $\overline{G}$ is also split and that both $G$ and $\overline{G}$ are chordal.

Conversely, suppose that $G$ and $\overline{G}$ are both chordal. Suppose $T$ is a minimum-diameter clique tree of $G$ and, arguing toward a contradiction, that the diameter of $T$ is at least three, and so $T$ is not a star. Then there exist vertices $Q_1$ and $Q_2$ that are the middle two vertices in an induced $P_4$ in $T$. Among the other neighbors of $Q_1$ in $T$ there must be a $Q_0$ for which there exists a vertex $v \in (Q_0 \cap Q_1)\backslash Q_2$, since otherwise all the neighbors (other than $Q_2$) of $Q_1$ in $T$ could be made adjacent to $Q_2$ instead of $Q_1$ and so create a clique tree for $G$ with smaller diameter than $T$. Let $u$ be a vertex in $Q_0 \backslash Q_1$. Similarly, there is a neighbor $Q_3 \neq Q_1$ of $Q_2$ and vertices $w \in (Q_2 \cap Q_3)\backslash Q_1$, and $x \in Q_3 \backslash Q_2$. Then $\{u, v, w, x\}$ induces a subgraph of $G$ with edge set $\{uv, wx\}$. But that would force an induced $C_4$ in $\overline{G}$, contradicting that $\overline{G}$ is chordal. □

**Exercise 2.26 (Földes & Hammer)** Show that a graph is a split graph if and only if none of its induced subgraphs is isomorphic to $2K_2$, $C_4$, or $C_5$.

The following corollary, from [McMorris & Shier, 1983], characterizes split graphs as intersection graphs. (Compare it with Exercise 2.3.) Recall that a *star* is a graph isomorphic to $K_{1,n}$ ($n \geq 0$); in other words, a tree $T$ with diameter at most two. A *substar* of a star is then simply a subtree of the star.

**Corollary 2.17 (McMorris & Shier)** *A graph is a split graph if and only if it is the intersection graph of a set of distinct substars of a star.*

**Proof.** First suppose $G$ is a split graph. Then $G$ is chordal by Theorem 2.16, and by its proof any minimum-diameter clique tree $T$ for $G$ has diameter less than three and so is a star. Then $G$ is the intersection graph of the substars $\{T_v : v \in V(G)\}$ of that star. If the diameter of $T$ is two, then these subtrees are distinct; if the diameter is one, then leaves can be added for each $v \in V(G)$ to produce distinct subtrees.

Conversely, suppose $G$ is the intersection graph of a set of distinct substars of a star $T$. We can suppose that $T$ is a clique tree for $G$ and that $Q \in V(T)$ is adjacent to all other vertices of $T$. Then $G$ is split, as shown by the partition of $V(G)$ into the complete subgraph $Q$ and the independent subset $I = V(G)\backslash Q$. □

**Exercise 2.27** Find an example of a graph that is not a split graph, yet is the intersection graph of a family of (not necessarily distinct) substars of a star. (What does Theorem 1.5 have to do with this?)

# Chapter 3

# Interval Graphs

An *interval graph* is defined to be any graph that is isomorphic to the intersection graph of a family of finite closed intervals of the real line, with each vertex $v$ corresponding to a closed interval $J_v$; the family of intervals is called an *interval representation* for the interval graph.

Interval graphs were first studied in [Hajós, 1957]. The standard references are section 3.4 of [Roberts, 1976] and Chapter 8 of [Golumbic, 1980].

**Example 3.1** The graph shown on the left in Figure 3.1 could have the representation $J_a = [1,4]$, $J_b = [1,1]$ (a single-point closed interval), $J_c = [1,2]$, $J_d = [2,3]$, $J_e = [3,4]$, and $J_f = [4,4]$. If you are squeamish about length-zero intervals, you could of course use $J_b = [1, 1.1]$ and $J_f = [3.1, 4]$ instead. You could also use all open intervals, instead of closed.

## 3.1 Definitions and Characterizations

Much as in Chapter 2 for subtrees of a tree, it is easy to see that we can equivalently define interval graphs using subpaths of a path and so talk about *path representations*. Since subpaths of a path satisfy the Helly condition, Lemma 1.11 can be used to show that every interval graph has a *clique path representation*, or *clique path* (paralleling Theorem 2.1 for trees).

**Example 3.1 (continued)** The graph shown in Figure 3.1 has the clique path $P$ shown there in which $P_a$ has length three, $P_b$ and $P_f$ have length zero, and $P_c$, $P_d$, and $P_e$ have length one.

Since paths are trees, interval graphs are chordal graphs, and so the cycle $C_4$ is a cheap example of a graph that is not an interval graph. A significantly different example would result from adding an edge $dg$ to the graph

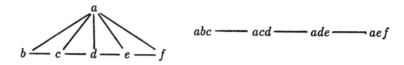

Figure 3.1: *An interval graph with a path representation.*

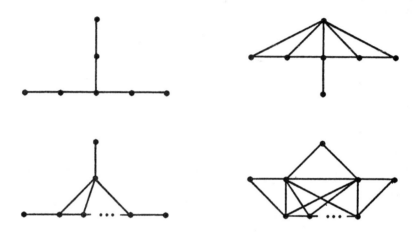

Figure 3.2: *Some graphs that are not interval graphs; each of the lower two has order at least six.*

on the left in Figure 3.1, producing the upper-right graph in Figure 3.2. (It might be instructive to attempt to find an interval representation or a path representation for that graph right now, instead of waiting for the characterizations just ahead in this chapter that show why it is impossible.) Indeed, none of the graphs shown in Figure 3.2 is an interval graph, while the graph in Figure 2.2 is an interval graph.

The following theorem consolidates the connection between interval and chordal graphs, extending a well-known theorem from [Fulkerson & Gross, 1965].

**Theorem 3.1** *A connected graph $G$ is an interval graph if and only if some maximum spanning tree of the weighted clique graph $K^w(G)$ is a clique path for $G$. Moreover, this is equivalent to every maximum spanning tree of the weighted clique graph $K^w(G)$ that is a path being a clique path for $G$.*

**Proof.** This is immediate from Theorem 2.3.                    □

**Corollary 3.2 (Fulkerson & Gross)** *A graph is an interval graph if and only if the edge clique cover of all maxcliques can be arranged into a clique path representation.*                    □

It is impractical to look for clique path representations by using Kruskal's algorithm to find all the maximum spanning trees and then trying to cull the nonpaths (indeed, being able to do that would be tantamount to solving the NP-hard problem of recognizing graphs that have hamiltonian paths). Yet interval graphs can be recognized efficiently. [Booth & Leuker, 1976] contains the classical, linear-time recognition algorithm, using the influential "PQ-tree" data structure that was introduced for that purpose; see also section 8.3 of [Golumbic, 1980]. [Simon, 1991], [Hsu & Ma, 1991], [Hsu, 1993], and [Corneil, Olariu, & Stewart, 1998] contain more recent recognition algorithms.

Theorem 3.1 relates to two prescient applications: Benzer's 1959 study of the fine structure of the gene, and Petrie's late nineteenth century work with archaeological seriation (see [Roberts, 1976] or [Golumbic, 1980]). Interval graphs *could* have been used in these contexts, but working with the appropriate incidence matrices as in Corollary 3.9 was quite sufficient. Section 3.4 contains "real" applications of interval graphs.

Three vertices form an *asteroidal triple* in a graph $G$ if, for each two, there exists a path containing those two but no neighbor of the third. For instance, the three vertices of degree one in the upper-left graph in Figure 3.2 form an asteroidal triple. Notice that no two vertices of an asteroidal triple can be adjacent. (Section 7.6 will discuss those graphs that have no asteroidal triples.)

**Exercise 3.1** Show that each of the graphs in Figure 3.2 has a unique asteroidal triple.

Theorem 3.3 is from [Lekkerkerker & Boland, 1962]. Recall from Chapter 2 that a graph is chordal if and only if contains no cycle $C_k$ having $k \geq 4$ as an induced subgraph.

**Theorem 3.3 (Lekkerkerker & Boland)** *A graph is an interval graph if and only if it is chordal and has no asteroidal triple.*

**Proof.** First suppose $G$ is an interval graph with a clique path $P$. As we have already observed, $P$ is also a clique tree and so $G$ is chordal by

Theorem 2.1. Suppose $u, v, w$ are pairwise nonadjacent and, without loss of generality, that $P_v$ is in between $P_u$ and $P_w$ along $P$. Since $P$ is a clique path, every path from $u$ to $w$ in $G$ will have to contain a neighbor of $v$, and so $u, v, w$ cannot be an asteroidal triple.

Conversely, suppose $G$ is a chordal graph and, among all clique trees for $G$, that $T$ has a minimum number of leaves and that number is at least three; we show that $G$ must then contain an asteroidal triple. Suppose $Q_1$, $Q_2$, and $Q_3$ are three different leaves of $T$ and let, respectively, $Q_1'$, $Q_2'$, and $Q_3'$ (not necessarily distinct) be their unique neighbors in $T$.

For each $i = 1, 2, 3$, choose $v_i \in V(Q_i)$ such that $v_i \notin V(Q_i')$. We show that $\{v_1, v_2, v_3\}$ is an asteroidal triple. Suppose rather, without loss of generality, arguing toward a contradiction, that every path in $G$ connecting $v_1$ and $v_3$ contains a neighbor of $v_2$. Not every edge of the path $T(Q_1, Q_3)$ in $T$ can contain a nonneighbor of $v_2$, since otherwise those nonneighbors could be used to induce a $v_1$-to-$v_3$ path in $G$ that contained no neighbor of $v_2$. Therefore, the path $T(Q_1, Q_3)$ in $T$ would contain some edge $Q^* Q^{**}$ with $Q^* \cap Q^{**}$ consisting entirely of neighbors of $v_2$, making $Q^* \cap Q^{**} \subseteq Q_2 \cap Q_2'$. Without loss of generality, suppose $Q^*$ is closer to $Q_2$ in $T$ than is $Q^{**}$. Then replacing edge $Q^* Q^{**}$ with a new edge $Q^{**} Q_2$ would create a clique tree for $G$ that has one fewer leaf than $T$, which is a contradiction.                  □

Using Theorem 3.3, [Lekkerkerker & Boland, 1962] proves that a graph is an interval graph if and only if it is chordal and contains none of the graphs in Figure 3.2 as an induced subgraph. [Harary & Kabell, 1984] contains a similar characterization of "infinite-interval graphs" in which the intervals are taken to be one- or two-way infinite intervals of the real line.

**Exercise 3.2** Let $G$ be a split graph. Show that $G$ is an interval graph if and only if $G$ contains none of the graphs in Figure 3.3 as an induced subgraph.

Before stating the next characterization of interval graphs, we need to review some terminology and results about directed graphs (digraphs). A digraph $D$ is defined to have a vertex set $V(D)$ and a set $A(D)$ of arcs, where $vw \in A(D)$ denotes an arc *from* vertex $v$ *to* vertex $w$. We assume that there are no multiple arcs (meaning that there are never two arcs from $v$ to $w$, although it is possible to have both $vw, wv \in A(D)$) and, in this chapter, no loops (meaning no $vv \in A(D)$). A digraph is *transitive* if, for $u, v, w \in V(D)$, $uv, vw \in A(D)$ with $u \neq w$ implies that $uw \in A(D)$. Given any graph $G$, an *orientation* of $G$ is a digraph formed by specifying a direction for each edge

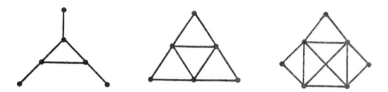

Figure 3.3: *Three split graphs that are not interval graphs.*

of $G$, producing an *oriented graph*. The orientation is a *transitive orientation* if the oriented graph is transitive.

**Exercise 3.3** Show that the cycle $C_4$ has a transitive orientation but that $C_5$ does not.

A *directed hamiltonian path* of a digraph is a directed path that includes every vertex. A *tournament* is an orientation $D$ of a complete graph; thus $u, v \in V(D)$ and $u \neq v$ imply that either $uv \in A(D)$ or $vu \in A(D)$ but not both. The following is from [Rédei, 1934].

**Lemma 3.4 (Rédei)** *Every tournament has a directed hamiltonian path.*

**Proof.** Suppose $D$ is a tournament. We argue by induction on $n = |V(D)|$, with the result trivial for $n \leq 2$. Suppose $n > 2$ and $v \in V(D)$. By induction hypothesis, the tournament $D - v$ has a directed hamiltonian path $v_1, v_2, \ldots, v_{n-1}$. If $vv_1 \in A(D)$, then $v, v_1, \ldots, v_{n-1}$ is a directed hamiltonian path in $D$. Otherwise, $v_1 v \in A(D)$ and we choose $i$ to be the largest integer for which $v_i v \in A(D)$. If $i = n$, then $v_1, \ldots, v_{n-1}, v$ is a directed hamiltonian path in $D$. If $i < n$, then $vv_{i+1} \in A(D)$, making $v_1, \ldots, v_i, v, v_{i+1}, \ldots, v_{n-1}$ a directed hamiltonian path in $D$. $\square$

**Exercise 3.4** Show that every transitive tournament of order $n$ has a unique vertex of each possible out-degree $0, \ldots, n-1$ and that taking these in order determines a directed hamiltonian path.

**Exercise 3.5** Show that every transitive tournament has a *unique* directed hamiltonian path.

Recall that, for a graph $G$, the *complement* of $G$, denoted $\overline{G}$, is the graph having $V(\overline{G}) = V(G)$ where, for any distinct vertices $u$ and $v$ of $G$, $uv \in E(\overline{G})$ if and only if $uv \notin E(G)$.

We are finally ready to state and prove the characterization from [Gilmore & Hoffman, 1964].

**Theorem 3.5 (Gilmore & Hoffman)** *A graph is an interval graph if and only if it does not contain $C_4$ as an induced subgraph and its complement has a transitive orientation.*

**Proof.** First suppose $G$ is an interval graph with clique path $P$ laid out horizontally. Since $G$ must be chordal, it cannot contain an induced $C_4$.

Form a oriented graph $\vec{G}$ by putting $uv \in A(\vec{G})$ if and only if $P_u$ is totally to the left of $P_v$ in $P$ (i.e., every vertex of $P_u$ is to the left of every vertex of $P_v$), noting that $uv \in A(\vec{G})$ implies $P_u \cap P_v = \emptyset$ and so $uv \notin E(G)$. It is easy to see that this is a transitive orientation of $\overline{G}$.

Conversely, suppose $G$ contains no induced $C_4$ and that $\overline{G}$ has a transitive orientation $\vec{G}$. We form a digraph $D$ whose vertices are precisely the maxcliques of $G$, with arcs as follows: For every two maxcliques $Q, Q'$ of $G$ pick $v \in Q$ and $v' \in Q'$ such that $vv' \notin E(G)$, and then put $QQ' \in A(D)$ if and only if $vv' \in A(\vec{G})$. Of course we must show that $D$ really is well defined. Arguing toward a contradiction, suppose that $u, v \in Q$ and $u', v' \in Q'$ where $uu', vv' \notin E(G)$ and $uu', v'v \in A(\vec{G})$. Observe that either $uv' \notin E(G)$ or $u'v \notin E(G)$, since the cycle $u, v, u', v', u$ cannot be induced in $G$. Without loss of generality, we suppose that $uv' \notin E(G)$. Thus either $uv' \in A(\vec{G})$ or $v'u \in A(\vec{G})$. If $uv' \in A(\vec{G})$, then $uv', v'v \in A(\vec{G})$ forces $uv \in A(\vec{G})$, since $\vec{G}$ is transitively oriented. But then $uv \in E(\overline{G})$, contradicting that $u$ and $v$ are in a common maxclique of $G$. A similar contradiction occurs if $v'u \in A(\vec{G})$. Thus $D$ is well defined.

It is easy to check that $D$ is transitive since $\vec{G}$ is transitive, so $D$ is a transitive tournament. By Lemma 3.4, $D$ has a directed hamiltonian path $P : Q_1, \ldots, Q_m$. We now show that $P$ is a clique path for $G$. Suppose that $v \in V(G)$ is in two nonadjacent vertices $Q, Q''$ of $P$ yet, arguing toward a contradiction, that $v \notin Q'$ for some $Q' \in V(P(Q, Q''))$. Without loss of generality, we can assume that $QQ'' \in A(D)$. Pick $u, w \in Q'$ such that $u \notin Q$ and $uv \notin E(G)$, while $w \notin Q''$ and $wv \notin E(G)$. Then $vu, wv \in A(\vec{G})$, so $u \neq w$ and, by transitivity, $wu \in A(\vec{G})$. But then $uw \notin E(G)$, contradicting that $u, w \in Q'$.                                                                $\square$

**Example 3.2** Figure 3.4 shows one possible transitive orientation of $\vec{G}$, where $G$ is the graph in Figure 3.1. Check that the only other transitive orientation is the reverse of this one. The corresponding digraph $D$ used in the preceding proof is also shown. Notice that the directed hamiltonian path in $D$ corresponds to the clique path in Figure 3.1.

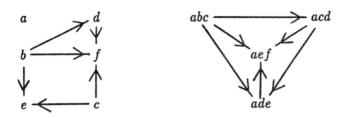

Figure 3.4: *A transitively oriented digraph to illustrate the proof of Theorem 3.5.*

[Opsut & Roberts, 1981] shows that the intersection number of an interval graph equals the number of maxcliques minus the number of isolated vertices.

## 3.2   Interval Hypergraphs

Continuing the discussion of tree hypergraphs in section 2.3, a hypergraph $(X, \mathcal{E})$ is an *interval hypergraph* if there is a path $P$ with $X = V(P)$ such that, for each $S_i \in \mathcal{E}$, there is a subpath $P_i$ of $P$ with $V(P_i) = S_i$.

**Exercise 3.6** Show that the hypergraph given in Exercise 2.17 is an interval hypergraph.

Precisely as in the easy direction of Theorem 2.7, every interval hypergraph must be a Helly hypergraph with an interval line graph. But, unlike what happened for tree hypergraphs, the following exercise shows that the converse fails.

**Exercise 3.7** Show that the hypergraph $(\{a, b, c, d\}, \mathcal{E})$ with $\mathcal{E} = \{\{a, c\}, \{b, c\}, \{c, d\}\}$ is a Helly hypergraph and its line graph $\Omega(\mathcal{E})$ is an interval graph, but no path $P$ exists as required in the definition of an interval hypergraph.

A *path in a hypergraph* $(X, \mathcal{E})$ is a sequence $v_0, S_1, v_1, S_2, \ldots, S_k, v_k$ where $S_1, \ldots, S_k$ are distinct edges, $v_0, \ldots, v_k$ are distinct vertices, and each $v_{i-1}v_i \in S_i$; such a path is said to *join* the vertices $v_0$ and $v_k$. The hypergraph $(X, \mathcal{E})$ is *connected* if every two vertices are joined by a path. A vertex $v$ is said to *lie between* vertices $u$ and $w$ in a hypergraph if every path in the hypergraph that joins $u$ and $w$ contains an edge that contains $v$. [Duchet, 1978, 1984]

contain proofs of the following characterization that is in the spirit of the Fulkerson–Gross result in Corollary 3.2.

**Theorem 3.6 (Duchet)** *A connected hypergraph is an interval hypergraph if and only if, for every three vertices, one of them lies between the other two.*

**Proof.** First, suppose $(X, \mathcal{E})$ is any connected hypergraph and there is a path $P$ with $X = V(P)$ for which each $S_i \in \mathcal{E}$ corresponds to a subpath $P_i$ of $P$ such that $V(P_i) = S_i$. Then a vertex $y$ lies between vertices $x$ and $z$ along $P$ if and only if $y$ lies between $x$ and $z$ in the hypergraph.

Conversely, suppose that $(X, \mathcal{E})$ satisfies the condition in the theorem— so for every three vertices, one of them lies between the other two. Choose a hypergraph $(X, \mathcal{E}^*)$ with $\mathcal{E} \subseteq \mathcal{E}^*$ such that, among hypergraphs satisfying the condition in the theorem, $\mathcal{E}^*$ is maximal. Let $\mathcal{E}'$ be the set of minimal edges $S$ of $\mathcal{E}^*$ for which $|S| \geq 2$.

Suppose $a$ and $b$ are distinct vertices in $S \in \mathcal{E}'$. If $S \neq \{a, b\}$, then $\{a, b\} \notin \mathcal{E}^*$ and so the hypergraph $(X, \mathcal{E}^* \cup \{\{a, b\}\})$ will not satisfy the condition in the theorem—$X$ contains $x, y, z$ and $(X, \mathcal{E}^* \cup \{\{a, b\}\})$ contains minimal-length paths

$$x, E_1, \ldots, E_p, y \text{ with } z \notin E_1 \cup \cdots \cup E_p,$$
$$y, E_1', \ldots, E_q', z \text{ with } x \notin E_1' \cup \cdots \cup E_q',$$
$$x, E_1'', \ldots, E_r'', z \text{ with } y \notin E_1'' \cup \cdots \cup E_r'',$$

where $\{a, b\} \in \{E_1, \ldots, E_p, E_1', \ldots, E_q', E_1'', \ldots, E_r''\}$. But if each occurrence of $\{a, b\}$ among the $E_i$'s, $E_i'$'s, and $E_i''$'s is replaced by $S \in \mathcal{E}^*$, then by the condition in the theorem one of $x, y, z$ will be between the other two; without loss of generality, say that $y$ is between $x$ and $z$. That means that $\{a, b\} = E_i''$ for some $1 \leq i \leq r$, and so $a \neq y \neq b$ and $y \in S$. Without loss of generality, using the minimality of the path $x, E_1'', \ldots, E_r'', z$, we can suppose that $a \in E_{i-1}''$ (or, possibly, $a = x$). By the assumed maximality of $\mathcal{E}^*$, we can assume that $A = E_1'' \cup \cdots \cup E_{i-1}'' \in \mathcal{E}^*$ (or, if $a = x$, that $A = \{a\} \in \mathcal{E}^*$). Thus there exists $A \in \mathcal{E}^*$ such that $a, x \in A$ and $b, y \notin A$. Similarly, there exists $B \in \mathcal{E}^*$ such that $b, z \in B$ and $a, y \notin B$. Again using the assumed maximality of $\mathcal{E}^*$, we can assume that $S \backslash A, S \backslash B \in \mathcal{E}^*$, and so $a$ and $b$ are connected by the path $a, S \backslash B, y, S \backslash A, b$ in $(X, \mathcal{E}^*)$.

Thus we have shown that every pair of vertices in an edge $S \in \mathcal{E}^*$ are linked by a path in $(X, \mathcal{E}^*)$ whose edges are subsets of $S$, and that every edge $S \in \mathcal{E}'$ has cardinality two, since otherwise $S \backslash B \in \mathcal{E}^*$ would contradict $S$'s assumed minimality. Therefore, $(X, \mathcal{E}')$ is a graph. The assumed maximality of $\mathcal{E}^*$ implies that $X \in \mathcal{E}^*$, so $(X, \mathcal{E}')$ is connected, and the condition

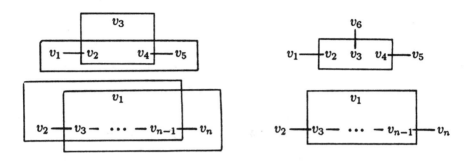

Figure 3.5: *The four noncycle forbidden subhypergraphs for an interval hypergraph.*

in the theorem implies that $(X, \mathcal{E}')$ is a path. Finally, every edge $S \in \mathcal{E}^*$ can be seen to be a connected subset of $(X, \mathcal{E}')$ by an inductive argument on $|S|$.  □

[Tucker, 1972] proves that a hypergraph is an interval hypergraph if and only if it contains none of the following five induced subhypergraphs $(\{v_1, \ldots, v_n\}, \mathcal{E})$ (see Figure 3.5 for the last four):
(1) $n \geq 3$ and $\mathcal{E} = \{\{v_1, v_2\}, \{v_2, v_3\}, \ldots, \{v_n, v_1\}\}$.
(2) $n = 5$ and $\mathcal{E} = \{\{v_1, v_2\}, \{v_2, v_3, v_4\}, \{v_4, v_5\}, \{v_1, v_2, v_4, v_5\}\}$.
(3) $n = 6$ and $\mathcal{E} = \{\{v_1, v_2\}, \{v_2, v_3, v_4\}, \{v_4, v_5\}, \{v_3, v_6\}\}$.
(4) $n \geq 4$ and $\mathcal{E} = \{\{v_2, v_3\}, \ldots, \{v_{n-1}, v_n\}, \{v_1, v_2, \ldots, v_{n-1}\},$
$\{v_1, v_3, \ldots, v_n\}\}$.
(5) $n \geq 4$ and $\mathcal{E} = \{\{v_2, v_3\}, \ldots, \{v_{n-1}, v_n\}, \{v_1, v_3, \ldots, v_{n-1}\}\}$.
[Trotter & Moore, 1976] gives a shorter proof, and [Duchet, 1984] contains a short proof using Theorem 3.6.

See [Lehel, 1983] and [Duchet, 1984, 1995] and their references for more on various sorts of representation of hypergraphs by intervals.

## 3.3 Proper Interval Graphs

A *proper interval graph* is the intersection graph of a family of closed intervals of the real line, none of which is *properly* contained in another. This is equivalent to being the intersection graph of a family of subpaths of a path, none of which is a proper subpath of another; such a path is called a *proper path representation* of the graph. For instance, Figure 3.6 shows a proper

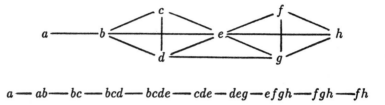

Figure 3.6: *A proper interval graph with a proper path representation.*

interval graph and a proper path representation for it. (Notice that we cannot have the vertices of the path just be the maxcliques, as was true for clique paths.) Proper interval graphs were introduced in [Roberts, 1969a] as "indifference graphs," for reasons we discuss in subsection 3.4.2; see also section 8.5 of [Golumbic, 1980]. They also were introduced in epidemiology as "time graphs"; see [Hedman, 1984].

**Exercise 3.8** Show that a proper interval graph cannot contain $K_{1,3}$ (the graph on the left in Figure 3.7) as an induced subgraph, and so that the graph in Figure 3.1 is not a proper interval graph.

**Exercise 3.9** Show that the clique graph of an interval graph must be a proper interval graph.

**Exercise 3.10** Show that every chordal graph has a "proper tree representation," meaning a tree representation $T$ in which no $T_v$ is properly contained in a $T_w$.

**Theorem 3.7 (Roberts)** *A graph is a proper interval graph if and only if it is an interval graph that does not contain an induced subgraph isomorphic to $K_{1,3}$.*

**Proof.** Exercise 3.8 gives the implication one way. For the converse, suppose $G$ has a clique path $P$ and contains no induced subgraph isomorphic to $K_{1,3}$. Suppose the subpath $P_v$ is properly contained in $P_w$. We first show that there cannot be vertices $Q^\ell, Q^r$ in $P_w$ with $P_v$ in between such that there are $x \in Q^\ell \backslash \{Q : Q \in P_v\}$ and $y \in Q^r \backslash \{Q : Q \in P_v\}$, since otherwise $v, w, x, y$ would induce a $K_{1,3}$ centered at $w$. Thus we can assume that $P_v$ and $P_w$ share one common end-vertex. Without loss of generality, assume

Figure 3.7: *Three graphs that are not proper interval graphs.*

$Q$ is their common left end-vertex. Now modify $P$ by inserting a new vertex $Q'$ just to the left of $Q$ with $Q' = Q \backslash \{u : Q$ is a left end-vertex of $P_u$ and $P_v$ is properly contained in $P_u\}$. After this modification, $P$ is still a path representation of $G$ but $P_v$ is no longer contained in $P_w$ (or in any of the $P_u$'s in the definition of $Q'$). Repeating this lengthens $P$ into a proper path representation of $G$. □

**Exercise 3.11** Use the construction in the proof of Theorem 3.7 to make a clique path for the graph in Figure 3.6 into a proper path representation.

**Exercise 3.12 (Roberts)** Show that a graph is a proper interval graph if and only if it contains no cycle of length greater than or equal to four and contains none of the graphs of Figure 3.7 as induced subgraphs.

[Wegner, 1967] and [Roberts, 1969a] define a *unit interval graph* to be the intersection graph of a family of closed intervals of the real line, all of which have the same length (which is often taken to be one). This is equivalent to being the intersection graph of a family of subpaths of a path, all of which have the same length; such a path is called a *unit path representation* of the graph. Since every unit path representation is a proper path representation, every unit interval graph is a proper interval graph. But the following theorem shows that much more is true.

**Theorem 3.8 (Roberts)** *A graph is a proper interval graph if and only if it is a unit interval graph.*

**Proof.** As we observed, the implication one way is immediate. To prove the converse, suppose $G$ is a proper interval graph. Arguing inductively, suppose that, for every proper subgraph $G'$ of $G$, every proper path representation of $G'$ can be made into a unit path representation of $G'$ by simply inserting duplicates of some of the vertices into the path. Suppose $P$ is a proper path representation for $G$ and pick a $v \in V(G)$ that occurs in an end-vertex of $P$. Obtain a proper path representation $P^+$ from $P$ by inserting duplicate vertices into $P$ so that removing all occurrences of $v$ from $P^+$ would leave a unit path representation of the subgraph induced by $V(G)\backslash\{v\}$. We can assume that $v$ still occurs in an end-vertex of $P^+$. Let $k = |V(P_w^+)|$ for each $w \neq v$. Then $k \geq |V(P_v^+)|$, since $P^+$ is still a proper path representation, and so adding $k - |V(P_v^+)|$ new vertices, each equal to $\{v\}$, to the end-vertex that contains $v$ will produce a unit path representation of $G$.                                                   □

[Jackowski, 1992] defines an *astral triple* in a graph as three vertices such that, for each two, there exists a path containing those two but not the third vertex that does not have two consecutive vertices that are neighbors of the third. For instance, the three vertices of degree one in the graph on the left in Figure 3.7 form an astral triple (but not an asteroidal triple). Paralleling the characterization of interval graphs in Theorem 3.3, Jackowski proves that a graph is a proper interval graph if and only if it contains no astral triple.

**Exercise 3.13 (Jackowski)** Show that every nonchordal graph contains an astral triple of vertices.

For any graph $G$ with vertices indexed by $\{1, \ldots, n\}$ and maxcliques indexed by $\{1, \ldots, m\}$, define the *maxclique-vertex matrix* $M(G)$ to be the $m \times n$ matrix with entry $m_{ij} = 1$ if the $i$th maxclique contains the $j$th vertex, and $m_{ij} = 0$ otherwise. For instance, the graph $G$ in Figure 3.6 has the maxclique-vertex matrix

$$M(G) = \begin{pmatrix} 1 & 1 & 0 & 0 & 0 & 0 & 0 & 0 \\ 0 & 0 & 0 & 1 & 1 & 0 & 1 & 0 \\ 0 & 1 & 1 & 1 & 1 & 0 & 0 & 0 \\ 0 & 0 & 0 & 0 & 1 & 1 & 1 & 1 \end{pmatrix},$$

where the columns correspond to the vertices in alphabetical order and the rows correspond to the maxcliques in the order $ab, deg, bcde, efgh$.

A matrix has the *consecutive ones property for columns* if its rows can be permuted so as to make all the 1 entries in each column consecutive. The *consecutive ones property for rows* is defined similarly. For instance, you can show that the above matrix has the consecutive ones property for columns by interchanging the second and third rows; it also has the consecutive ones property for rows. The following merely rephrases Corollary 3.2.

**Corollary 3.9** *A graph $G$ is an interval graph if and only if $M(G)$ has the consecutive ones property for columns.*                              □

The following result seems to have been first stated in this form in [Deogun & Gopalakrishnan, to appear], in connection with the application in section 3.4.3, although all the pieces were certainly in [Roberts, 1968]. Section 7.1 contains more about variations of the consecutive ones properties.

**Theorem 3.10** *A graph $G$ is a proper interval graph if and only if $M(G)$ has the consecutive ones property for both rows and columns.*

**Proof.** First suppose $G$ has a proper path representation $P$. Corollary 3.9 shows that $M(G)$ has the consecutive ones property for columns. Let $v_1, \ldots, v_n$ be the vertices in the order of their leftmost appearance along $P$. Suppose $i < j < k$ and $v_i v_k \in E(G)$. Then $P_{v_i}$ (the subpath of $P$ corresponding to $v_i$) will intersect $P_{v_k}$ and so also $P_{v_j}$, ensuring that $v_i v_j \in E(G)$. Because $P_{v_j}$ cannot be properly contained in $P_{v_i}$, path $P_{v_j}$ will have to intersect $P_{v_k}$, ensuring that $v_j v_k \in E(G)$. Thus, each maxclique of $G$ will correspond to consecutive vertices in the ordering $v_1, \ldots, v_n$, and this means that $M(G)$ has the consecutive ones property for rows.

Conversely, suppose $M(G)$ has the consecutive ones property for both columns and rows. The former of these implies that $G$ is an interval by Corollary 3.9. The latter implies that $G$ does not contain $K_{1,3}$ as an induced subgraph, and so $G$ is a proper interval graph by Exercise 3.7.   □

Sections 3.4.2, 7.1, and 7.2 contain other characterizations of proper interval graphs, and [Gutierrez & Oubiña, 1996] contains various order-theoretic characterizations. [Gutierrez & Oubiña, 1995] shows that every proper interval graph satisfies

$$|V(G)| \geq 2c(G) - c(K(G)),$$

where $K(\cdot)$ is the clique graph operator from section 1.4 and $c(\cdot)$ counts the number of maxcliques and then investigates the graphs for which equality holds.

See [Corneil, Kim, Natarajan, Olariu, & Sprague, 1995], [Hell & Huang, 1995], and [de Figueiredo, Meidanis, & de Mello, 1995] for recognition algorithms for proper interval graphs, and [Hell & Huang, 1995] and [Deng, Hell, & Huang, 1996] for representation algorithms.

[Leibowitz, Assman, & Peck, 1982] generalizes the notion of a unit interval graph by defining the "interval count" of an interval graph to be the minimum number of different lengths of intervals needed in an interval representation. [Skrien, 1984] characterizes graphs that have interval count two, where one of the allowed lengths is zero—in other words, the intersection graphs of points and unit intervals; section 5.2 will show that "threshold" graphs are of this type.

[Pe'er & Shamir, 1995] investigates a host of restrictions on interval graphs, including bounding the maximum lengths of intervals.

## 3.4   Some Applications of Interval Graphs

Each of the following subsections is merely a brief sketch of one application of interval graphs. As we did in section 2.4, we have selected applications that make essential use of the intersection definition of interval graphs, rather than other important applications that involve interval graphs. One example of the latter involves on-line coloring algorithms: a graph is presented one vertex at a time, along with its neighbors among earlier vertices, and the graph is to be properly colored with as few colors as possible. This is a highly practical problem in many contexts, dynamic storage problems for one. Papers such as [Gyárfás & Lehel, 1988], [Ślusarek, 1989], [Kierstead, 1991] and [Kierstead & Qin, 1995] contain results for interval graphs; the first of these also studies proper interval graphs, while [Ślusarek, 1995] and [Marathe, Hunt, & Ravi, 1996] study circular-arc graphs (section 7.1).

### 3.4.1   Applications to Biology

Probably the first paper on interval graphs was an application in biology. Although stated in terms of incidence matrices rather than graphs, the question in [Benzer, 1959] was whether certain fragment overlap data on the DNA making up a bacterial gene was consistent with the gene having a linear structure—in other words, whether the graph constructed from the data was an interval graph. Of course today we know that the gene is indeed a linear arrangement and, as mentioned in section 2.4.1, DNA strands are sequences (words) built from the four letter alphabet {A,C,G,T}.

One of the problems involving DNA is to try to assemble subsequences involving possible overlaps into longer sequences. Certainly one would expect interval graphs and their variants to be useful, and indeed this has been the case. [Jungck, Dick, & Dick, 1982] is a very readable introductory paper. See [Fellows, Hallett, & Wareham, 1993], [Goldberg, Golumbic, Kaplan, & Shamir, 1995], and [Nicholson, 1995] for more recent views. [Mirkin

& Rodin, 1984] and [Waterman, 1995] are two excellent books with in-depth treatment.

We now briefly present some oversimplified background relevant to what is called *physical mapping* of DNA. Sequence fragments called *clones* are obtained from an unknown DNA sequence and form a "clone library." Experiments are then carried out that can decide if a very short molecule, called a *probe*, overlaps each clone. The goal is to reconstruct the placement of the clones along the DNA sequence, the sequence having been destroyed during the construction of the clone library. If only some of the clones are used as probes, then the overlap information is not available between clones that are not probes. [Zhang, to appear] introduced the following generalization of interval graphs to deal with this situation.

A graph $G$ is a *probe interval graph* if $V(G)$ can be partitioned into subsets $P$ and $N$ (corresponding to the probes and nonprobes) and each $v \in V(G)$ can be assigned to an interval $I_v$ such that $uv \in E(G)$ if and only if both $I_u \cap I_v \neq \emptyset$ and at least one of $u$ and $v$ is in $P$. Interval graphs are simply probe interval graphs with $N = \emptyset$.

**Exercise 3.14** Show that, although $C_4$ and the graph in the middle of Figure 3.7 are not interval graphs, they are both probe interval graphs.

Results on probe interval graphs and their variants can be found in [Zhang, to appear], [McMorris, Wang, & Zhang, to appear], [Wan, Lee, Wang, & Zhang, to appear] and [Sheng, Wang, & Zhang, to appear]; also see [Atkins & Middendorf, 1996]. The following result considerably restricts the possible structure of probe interval graphs; the graphs described therein are the *weakly chordal graphs*, that are discussed further in section 7.3.

**Theorem 3.11 (McMorris, Wang, & Zhang)** *Neither a probe interval graph nor its complement can contain an induced cycle of length greater than or equal to five.*

**Proof.** Suppose $G$ is a probe interval graph with respect to the partition $V(G) = P \cup N$, with $I_v$ the interval assigned to each $v \in V(G)$. Let $G^*$ be defined precisely the same as $G$ except with $uv \in E(G^*)$ if and only if $I_u \cap I_v \neq \emptyset$ (*without* the addition $\{u, v\} \cap P \neq \emptyset$ assumption). Clearly $G^*$ is an interval graph. Suppose $C$ is an induced cycle of $G$ of length at least four and $u, v \in P$ are adjacent along $C$. Since the only edges in $E(G^*) \setminus E(G)$ are between vertices of $N$, $u$ and $v$ have the same neighborhoods in $G$ and $G^*$. But then some subset of the vertices of $C$ would induce a chordless cycle containing $u$ and $v$ of length at least four in $G^*$, contradicting that $G^*$ is an interval (and so chordal) graph. Therefore, vertices of $P$ and $N$ alternate on $C$, and so $G$ has no induced cycles of odd length greater than three.

Now suppose $C$ is an induced cycle of $G$ of even length at least six. Let $x, y, z \in P$ be nonadjacent vertices along $C$, and let paths $P_1, P_2$, and $P_3$ be, respectively, the segments of $C$ between $x$ and $y$, between $y$ and $z$, and between $z$ and $x$; thus $z$ is not adjacent to any vertex on $P_1$, $x$ is not adjacent to any vertex on $P_2$, and $y$ is not adjacent to any vertex on $P_3$. Since $x, y, z \in P$, they have no new neighbors in $G^*$ and so form an asteroidal triple in $G^*$, again contradicting that $G^*$ is an interval graph. Therefore, a probe interval graph has no induced cycles of length greater than or equal to five.

To show that $G$ contains no complement of an induced cycle of length larger than four, first notice that the complement of an induced cycle of length five would also be an induced cycle of length five, which we now know is impossible in $G$. Therefore, we only need to show that $G$ contains no complement of an induced cycle of length six or more. Suppose to the contrary that $\{v_1, \ldots, v_n\}$ induces a complement of an induced cycle in $G$ with $n \geq 6$, where $v_1 v_n \notin E(G)$ and each $v_i v_{i+1} \notin E(G)$, while all other $v_i v_j$'s are in $E(G)$. Then $v_1, v_4, v_2, v_5, v_1$ will be an induced cycle in $G$, and so we can assume without loss of generality that $v_1, v_2 \in P$ and $v_4, v_5 \in N$. Therefore, $v_3 v_5 \in E(G)$ implies that $v_3 \in P$ and $v_4 v_6 \in E(G)$ implies that $v_6 \in P$. But then $v_2, v_5, v_3, v_6, v_2$ would be a length-four chordless cycle in $G$ having three vertices from $P$, contradicting vertices from $P$ and $N$ alternating around induced cycles in $G$.                                          $\square$

**Exercise 3.15 (Zhang)** Show that "enhancing" a probe interval graph by adding edges between pairs of nonprobes that have two nonadjacent probes as common neighbors produces a chordal graph.

Also related to DNA matters, [Bodlaender & de Fluiter, 1996] discusses a "chromatic interval completion problem," paralleling the chromatic chordal completion problem—inserting edges so as to make an interval graph—in section 2.4.1. Going the other way, [Wang, 1994] discusses removing edges from a bipartite graph so as to leave an interval graph.

### 3.4.2  Applications to Psychology

While classical theories of measurement are based on the physical sciences, much work has also been done on notions of measurement that are more suitable for the social sciences. Much of this work has been within the context of psychology. Our discussion below is based on [Roberts, 1976, 1978b]. [Roberts, 1979] is a related treatment of various aspects of "measurement theory."

Suppose $A$ is a set of alternatives, such as types of cars or food products, and a person has preferences among the elements of $A$. [Luce, 1956] motivates seeking a real-valued function $f$ on $A$ such that, for $a, b \in A$, preferring alternative $a$ to alternative $b$ implies $f(a) > f(b) + \delta$, where the positive constant $\delta$ represents a *threshold* or *just noticeable difference* between alternatives.

Define a binary relation $R$ on a finite set $A$ to be an *interval order* if it satisfies the two following axioms.

**Axiom 1:** *For all $a \in A$, not $aRa$.*

**Axiom 2:** *For all $a, b, c, d \in A$, if $aRb$ and $cRd$, then either $aRd$ or $cRb$.*

**Exercise 3.16** Suppose $f$ is a real-valued function defined on $A$, and $aRb$ is defined to mean that $f(a) > f(b) + \delta$, where $\delta$ is a positive constant. Show that $R$ is an interval order on $A$.

For any binary relation $R$ on any finite set $A$, define the graph $G(R)$ to have $V(G(R)) = A$ with $ab \in E(G(R))$ if and only if neither $aRb$ nor $bRa$; edges thereby correspond to "indifference" with respect to $R$. (Warning: An "indifference graph" per se is defined somewhat differently and is equivalent to being a proper interval graph.) The following result is from [Fishburn, 1970a, 1970b].

**Proposition 3.12 (Fishburn)** *A binary relation $R$ on a finite set $A$ is an interval order on $A$ if and only if $R$ is transitive and $G(R)$ is an interval graph.*

**Proof sketch.** First suppose $R$ satisfies Axioms 1 and 2; transitivity follows directly. Theorem 3.5 shows that $G = G(R)$ is an interval graph as follows: $G$ cannot contain an induced cycle $a, b, c, d, a$ since $ac, bd \notin E(G)$ would imply both ($aRc$ or $cRa$) and ($bRd$ or $dRb$), and each of the four possible cases would lead to a contradiction using Axiom 2; and $R$ is itself a transitive orientation of $\overline{G}$.

Conversely, suppose $R$ is transitive and $G = G(R)$ is an interval graph. Axiom 1 follows from $G$ being loopless. Arguing toward a contradiction with Axiom 2, suppose $aRb$ and $cRd$, so $ab, cd \notin E(G)$, yet neither $aRd$ nor $cRb$. Then by transitivity, neither $dRa$ nor $bRc$, so $ad, bc \in E(G)$. Transitivity similarly shows $ac, bd \in E(G)$, producing an induced cycle $a, d, b, c, a$, contradicting Theorem 3.5.                                           □

**Exercise 3.17** Discuss whether or not our proof of Proposition 3.12 actually shows something stronger: that a binary relation $R$ on a finite set $A$ is an interval order on $A$ if and only if $R$ is transitive and $G(R)$ contains no induced $C_4$.

Define a binary relation $R$ on a finite set $A$ to be a *semiorder* if it satisfies Axioms 1 and 2 and also the following axiom.

**Axiom 3:**  *For all $a, b, c, d \in A$, if $aRb$ and $bRc$, then either $aRd$ or $dRc$.*

**Exercise 3.18** Suppose $f$, $\delta$, $A$, and $R$ are as in Exercise 3.16. Show that $R$ is a semiorder of $A$.

We state the following result of [Roberts, 1969a, 1971], as stated in [Fishburn, 1985], without proof.

**Proposition 3.13 (Roberts)** *A binary relation $R$ on a finite set $A$ is a semiorder on $A$ if and only if $R$ is transitive and $G(R)$ is a proper interval graph.*

Again with [Luce, 1956] as motivation, [Roberts, 1971] defines a graph $G$ to be *representable by just noticeable differences* if, for each $v \in V(G)$, there exists a real number $r_v$ contained in a closed interval $J_v$ of the real line such that $uv \in E(G)$ if and only if $r_u \in J_v$ (or, equivalently, $r_v \in J_u$). (Compare this with the concept of "catch graphs" in section 7.2.) While we state the following result without proof, Exercise 3.19 will be a simpler special case.

**Proposition 3.14 (Roberts)** *A graph is representable by just noticeable differences if and only if it is a proper interval graph.*

**Exercise 3.19 (Roberts)** Show that a graph $G$ is a proper interval graph if and only if, for each $v \in V(G)$, there exists a real number $r_v$ and a closed *unit* interval $J_v$ centered at $r_v$ such that $uv \in E(G)$ if and only if $r_u \in J_v$.

More general applications to psychology involve general seriation problems. For instance in developmental psychology, [Coombs & Smith, 1973] studies whether psychological "traits" could correspond to chronological intervals—interval graph models would clearly be useful here. [Hubert, 1974] surveys the role of interval and proper interval graphs in seriation problems in psychology. [Troxell, 1995] also considers proper interval graphs.

### 3.4.3 Applications to Computing

Applications of interval graphs tend to be examples of general "seriation" problems, determining whether certain data or objects are compatible with arrangement in a linear pattern. Such examples are often scheduling problems, with the linear dimension corresponding to time. A common example involves a graph having university courses as vertices, with two vertices adjacent if and only if the courses overlap in time of day and so cannot be assigned a common room. Such a graph will be an interval graph, and finding the minimum number of rooms needed corresponds to finding the graph's chromatic number, a problem that is *much* easier for interval graphs than in general. (Making a hard—in this case NP-complete—problem tractable is an important role of interval graphs in computing, much as we mentioned for chordal graphs at the end of subsection 2.4.2; [Olariu, Schwing, & Zhang, 1995] is an up-to-date discussion.) [Kendall, 1969] contains another well-known "seriation in time" problem to which interval graphs are applicable, in this case to archaeology.

There are many applications to computing in which the seriation is not with respect to time. [Golumbic, 1984] gives one interesting example, and there are others in section 8.4 of [Golumbic, 1980]. Our discussion is in terms of the widely studied topic of consecutive retrieval file organization. The original idea appeared in [Ghosh, 1972], with [Eswaran, 1975] linking it to interval graphs. [Ghosh, Kambayashi, & Lipski, 1983] is a collection of articles on this subject, with [Lipski, 1983] listing almost 200 references on consecutive retrieval and interval graphs.

Suppose $\mathcal{R}$ is a set of *records* (files) and $\mathcal{Q}$ is a set of *queries*, each linked to a particular set of relevant records so that each query $Q_i \in \mathcal{Q}$ can be identified with a subset of $\mathcal{R}$. Such $\mathcal{R}$ and $\mathcal{Q}$ are said to satisfy the *consecutive retrieval property* if the records relevant to each query can be stored consecutively in linear storage without repeating records.

**Example 3.3** Suppose A, B, C, D, E, F, G, H, I are nine records with $Q_1 = \{A, B, C\}$, $Q_2 = \{D, E, F, G\}$, $Q_3 = \{C, D, H, I\}$, $Q_4 = \{D, I\}$, and $Q_5 =$

$\{C, D, E, H, I\}$. Then one way to satisfy the consecutive retrieval property is shown by the linear arrangement

$$A - B - C - H - I - D - E - F - G.$$

Clearly $\mathcal{R}$ and $\mathcal{Q}$ satisfy the consecutive retrieval property if and only if $H = (\mathcal{R}, \mathcal{Q})$ is an interval hypergraph. As we observed in section 3.2, this means that if the records of $\mathcal{R}$ can be arranged so as to satisfy the consecutive retrieval property with respect to the queries in $\mathcal{Q}$, then the line graph $\Omega(\mathcal{Q})$ of $H$ must be an interval graph. But, as we also observed in section 3.2, the converse fails. This is a subtle, but important, point that can cause confusion in various seriation applications. The subtlety is shown by its incorrect inclusion as a theorem in [Ghosh, 1977] and its removal from the second edition, [Ghosh, 1986]. That $\Omega(\mathcal{Q})$ is an interval graph corresponds to $\mathcal{Q}$ being representable by intervals, not to the arrangement of the members of $\mathcal{R}$. The hypergraphs of Example 2.6 and Exercise 3.7 can both be thought of as examples of $\mathcal{R}$ and $\mathcal{Q}$ where $\Omega(\mathcal{Q})$ is an interval graph, yet the records cannot be linearly arranged so as to satisfy the consecutive retrieval property. [Deogun & Gopalakrishnan, to appear] proves the following.

**Proposition 3.15 (Deogun & Gopalakrishnan)** *Suppose the hypergraph $H = (\mathcal{R}, \mathcal{Q})$ is such that there are not two $Q_i, Q_j \in \mathcal{Q}$ with $Q_i \subseteq Q_j$. Then the records in $\mathcal{R}$ can be arranged so as to satisfy the consecutive retrieval property with respect to $\mathcal{Q}$ if and only if the line graph of $H$ is isomorphic to the clique graph of the line graph of the dual hypergraph $H^*$ and this clique graph is a proper interval graph.*

The consecutive retrieval property has been generalized in various ways, many of them in [Ghosh, Kambayashi, & Lipski, 1983]. In particular, [Tanaka, 1983] investigates replacing linear storage with storage on trees, thereby replacing interval hypergraphs with tree hypergraphs.

# Chapter 4

# Competition Graphs

This chapter considers intersection graphs of various sorts of neighborhoods in graphs and digraphs, the most studied of which are the "competition graphs" in section 4.2. But, in a generic sense, they all can be thought of in terms of "competition."

The development of these topics differs from that of chordal and interval graphs in that they are intersection graphs of the set of *all* subgraphs (neighborhoods) of a certain sort, rather than an arbitrary multiset of them. They resemble clique and line graphs in this regard. In particular, each of these topics has an associated graph operator that is discussed more thoroughly in [Prisner, 1995].

## 4.1 Neighborhood Graphs

Recall that for any graph $G$ and any $v \in V(G)$, the *open neighborhood* of $v$, denoted $N_G(v)$, is the subgraph induced by $\{u : uv \in E(G)\}$. The *closed neighborhood* of $v$ in $G$, denoted $N_G[v]$, is the subgraph induced by $N_G(v) \cup \{v\}$. We write $N(v)$ and $N[v]$ when $G$ is clear from the context.

### 4.1.1 Squared Graphs

For any graph $G$, the *square* of $G$, denoted $G^2$, has the same vertices as $G$, with two vertices $u$ and $v$ adjacent if and only if $d(u, v) \leq 2$ in $G$, where $d(u, v)$ denotes the usual graph distance; this can be thought of as saying that $u$ and $v$ are close enough to "compete" in some sense. A graph $G$ is a *squared graph* if $G \cong H^2$ for some graph $H$.

**Example 4.1** The graph $G$ in Figure 4.1 is the square of the graph $H$.

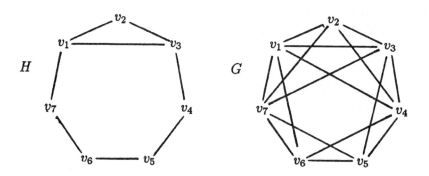

Figure 4.1: *A graph H and its square $G = H^2$.*

**Exercise 4.1** Describe the relationship between the incidence matrices of $G$ and $G^2$.

**Theorem 4.1** *A graph is a squared graph if and only if it is the intersection graph of all the closed neighborhoods of the vertices of some graph.*

**Proof.** This follows immediately from the observation that, for $u \neq v$, $N[u] \cap N[v] \neq \emptyset$ if and only if $d(u, v) \leq 2$. $\qquad\qquad\square$

Squared graphs originated in [Harary & Ross, 1960], where the squares of trees were characterized. Squared graphs in general were characterized in [Mukhopadhyay, 1967]. Notice how "$v \in N[v]$" translates into condition (1) on an edge clique cover in the following theorem, and "$u \in N[v]$ if and only if $v \in N[u]$" translates into condition (2).

**Theorem 4.2 (Mukhopadhyay)** *A graph $G$ with $V(G) = \{v_1, \ldots, v_n\}$ is a squared graph if and only if $G$ has an edge clique cover $\mathcal{E} = \{Q_1, \ldots, Q_n\}$ such that both the following hold:*
 (1) *for every $i$, $v_i \in Q_i$;*
 (2) *for every $i \neq j$, $v_i \in Q_j$ if and only if $v_j \in Q_i$.*

**Proof.** First suppose $G$ has vertex set $\{v_1, \ldots, v_n\}$ and edge clique cover $\mathcal{E} = \{Q_1, \ldots, Q_n\}$ satisfying conditions (1) and (2). Put $\mathcal{F} = \mathcal{F}(\mathcal{E})$, the dual set representation of $G$ determined from $\mathcal{E}$. Thus $G \cong \Omega(\mathcal{F})$ where, in the definition of $\mathcal{F}(\mathcal{E})$, $\mathcal{F} = \{S_1, \ldots, S_n\}$ and each $S_i = \{j : v_i \in Q_j\}$.

Define a graph $H$ on $V(H) = \{1, \ldots, n\}$ where $jk \in E(G)$ if and only if $j \neq k$ and $v_k \in Q_j$, noting that adjacency is indeed symmetric by (2). Fix

$k \in V(H)$. Then $j \in N_H[k]$ if and only if either $jk \in E(H)$ or $j = k$. So by (1), $j \in N_H[k]$ is equivalent to $v_k \in Q_j$, and so to $j \in S_k$. Thus each $S_k = N_H[k]$, and so $G$ is a squared graph by Theorem 4.1.

Conversely, suppose $G \cong H^2$ where for convenience we assume $V(G) = V(H^2) = \{v_1, \ldots, v_n\}$ with each $v_i \in V(G)$ corresponding to $v_i \in V(H^2)$ under the isomorphism. Thus $G \cong \Omega(\mathcal{F})$ where $\mathcal{F} = \{S_1, \ldots, S_n\}$ and each $S_i = N_H[v_i]$. Put $\mathcal{E} = \mathcal{E}(\mathcal{F})$, the dual edge clique cover of $G$ determined by $\mathcal{E}$; thus $\mathcal{E} = \{Q_1, \ldots, Q_n\}$ where, in the definition of $\mathcal{E}(\mathcal{F})$, each

$$Q_j = G_{v_j} = \{v_i : v_j \in S_i\} = \{v_i : v_j \in N_H[v_i]\} = N_H[v_j].$$

Condition (1) holds since each $v_i \in N_H[v_i]$, and (2) holds since $v_i \in N_H[v_j]$ is equivalent to $v_j \in N_H[v_i]$. □

**Example 4.1 (continued)** For the squared graph $G$ in Figure 4.1, $Q_1 = \{v_1, v_2, v_3, v_7\}$, $Q_2 = \{v_1, v_2, v_3\}$, $Q_3 = \{v_1, v_2, v_3, v_4\}$, $Q_4 = \{v_3, v_4, v_5\}$, $Q_5 = \{v_4, v_5, v_6\}$, $Q_6 = \{v_5, v_6, v_7\}$, $Q_7 = \{v_1, v_6, v_7\}$ form an edge clique cover as described in Theorem 4.2.

### 4.1.2 Two-Step Graphs

For any graph $G$, the *two-step graph* (or *two-path graph*) of $G$, denoted $G_2$, has the same vertices as $G$, with two vertices $u$ and $v$ adjacent if and only if there is a path of length exactly two connecting $u$ and $v$ in $G$. A graph $G$ is a two-step graph if $G \cong H_2$ for some graph $H$.

**Example 4.2** The graph $G$ in Figure 4.2 is the two-step graph of the graph $H$.

**Theorem 4.3** *A graph is a two-step graph if and only if it is the intersection graph of all the open neighborhoods of the vertices of some graph.*

**Proof.** This follows immediately from the observation that, for $u \neq v$, $N(u) \cap N(v) \neq \emptyset$ if and only if there is a path of length two connecting $u$ and $v$. □

[Escalante, Montejano, & Rojano, 1974] is a good reference on two-step graphs, which were characterized in [Acharya & Vartak, 1973], noting the "striking similarity" with squared graphs.

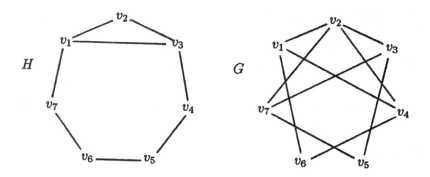

Figure 4.2: *A graph H and its two-step graph $G = H_2$.*

**Theorem 4.4 (Acharya & Vartak)** *A graph G with $V(G) = \{v_1,$ $\ldots, v_n\}$ is a two-step graph if and only if G has an edge clique cover $\mathcal{E} = \{Q_1, \ldots, Q_n\}$ such that both the following hold:*
  (1) *for every i, $v_i \notin Q_i$;*
  (2) *for every $i \neq j$, $v_i \in Q_j$ if and only if $v_j \in Q_i$.*

**Proof.** This can be proved by a minor modification of the proof of Theorem 4.2.                                                      □

**Exercise 4.2** Find the edge clique cover $\mathcal{E}$ as in Theorem 4.4 for the two-step graph $G$ in Figure 4.2.

**Exercise 4.3** Verify the details in the proof of Theorem 4.4.

## 4.2  Competition Graphs

Recall that a digraph $D$ can be defined to have a finite vertex set $V(D)$ and a set $A(D)$ of arcs, where $vw \in A(D)$ denotes an arc *from* vertex $v$ *to* vertex $w$. We assume that there are no multiple arcs (meaning that there are never two arcs from $v$ to $w$, although it is possible to have both $vw, wv \in A(D)$). In this chapter, we will sometimes allow loops (meaning an arc $vv$). For each $v \in A(D)$, define the *out-neighborhood* of $v$ in $D$, denoted $N_D^+(v)$ or $N^+(v)$, to be the subdigraph of $D$ induced by $\{w : vw \in A(D)\}$. (Notice that $v \in N^+(v)$ if and only if $D$ contains a loop at $v$.) A *sink* of $D$ is a vertex $v \in V(D)$ such that $N^+(v) = \emptyset$. Similarly, the *in-neighborhood* of $v$ in $D$, denoted $N_D^-(v)$ or $N^-(v)$, denotes the subdigraph of $D$ induced by $\{w : wv \in A(D)\}$. A *source* of $D$ is a vertex $v \in V(D)$ such that $N^-(v) = \emptyset$.

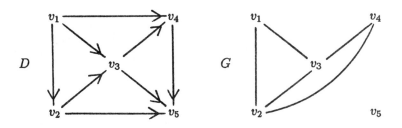

Figure 4.3: *A digraph D and its competition graph G = C(D).*

Competition graphs were introduced by Joel Cohen in 1968 in the context of a *food web* (an acyclic digraph) $D$ whose vertices are *species* with $vw \in A(D)$ whenever species $v$ feeds on species $w$. Competition graphs are sometimes called *niche overlap graphs* or *consumer graphs* for this reason. (Warning: Graph theorists direct arcs from "predators" toward "prey," but biologists use the opposite directions for the arcs when they draw foodwebs.) [Cohen, 1978] and [Roberts, 1976, 1978a] contain much more information. This biological motivation also explains the frequent restriction to acyclic (and so, automatically, to loopless) digraphs. [Lundgren, 1989] is a recent survey.

The *competition graph* $C(D)$ of a digraph $D$ has vertex set $V(D)$ and edges so as to make $C(D) \cong \Omega(\{N^+(v) : v \in V(D)\})$. Notice that a sink in $D$ will be an isolated vertex in $C(D)$, and that $uv \in E(C(D))$ if and only if there is some $w \in V(D)$ such that both $uw, vw \in A(D)$—which means if and only if $u$ and $v$ are in some common $N^-(w)$. In other words, two species are adjacent in the competition graph if and only if they compete for (both feed on) some common prey.

**Example 4.3** The graph $G$ in Figure 4.3 is the competition graph of the acyclic digraph $D$ shown there. For instance, $v_1 v_3 \in E(G)$ because $N^+(v_1) \cap N^+(v_3) = \{v_2, v_3, v_4\} \cap \{v_4, v_5\} \neq \emptyset$—species $v_1$ and $v_3$ compete for species $v_4$—while $v_1 v_4 \notin E(G)$ because $N^+(v_1) \cap N^+(v_4) = \{v_2, v_3, v_4\} \cap \{v_5\} = \emptyset$—species $v_1$ and $v_4$ do not both feed on a common species. Vertex $v_5$ is isolated in $G$, since it is a sink in $D$. The vertex labels used in Figure 4.3 are as described in the following lemma.

**Lemma 4.5** *If a digraph $D$ is acyclic, then $V(D)$ can be labeled as $\{v_1, \ldots, v_n\}$ so that $v_i v_j \in A(D)$ implies that $i < j$.*

**Proof.** Suppose $D$ is acyclic. If every vertex of $D$ had positive in-degree, then moving backward along arcs would eventually determine a directed cycle. Thus some vertex must have in-degree zero; label it $v_1$ and remove it from the digraph. Repeat this procedure—each time labeling a vertex with in-degree zero using the next available label—until all vertices are labeled.

Arguing toward a contradiction, suppose $v_i v_j \in A(D)$ yet $i > j$. When $v_i$ becomes labeled, $v_j$ will already have become labeled within a digraph containing the vertex that becomes $v_i$. But this means that $v_j$ had positive in-degree when it became labeled, which is a contradiction.                    □

**Exercise 4.4** Prove the converse of Lemma 4.5.

The basic characterization of competition graphs of acyclic digraphs is in both [Dutton & Brigham, 1983] and [Lundgren & Maybee, 1983a]. Recall that the definition of an edge clique cover $\mathcal{E}$ in section 1.1 allows any $Q_i \in \mathcal{E}$ to be the null subgraph.

**Theorem 4.6   (Dutton & Brigham and Lundgren & Maybee)** *A graph $G$ is a competition graph of an acyclic digraph if and only if $V(G)$ can be labeled as $\{v_1, \ldots, v_n\}$ and $G$ has an edge clique cover $\mathcal{E} = \{Q_1, \ldots, Q_n\}$ such that $v_i \in Q_j$ implies $i < j$.*

**Proof.** First suppose $G$ has vertex set $\{v_1, \ldots, v_n\}$ and edge clique cover $\mathcal{E} = \{Q_1, \ldots, Q_n\}$ such that $v_i \in Q_j$ implies $i < j$. Define a digraph $D$ with $V(D) = V(G)$, where $v_i v_j \in A(D)$ if and only if $v_i \in Q_j$, noting that $D$ is acyclic by Exercise 4.4. Then $v_k v_i \in E(G)$ if and only if $v_k, v_i \in Q_j$ for some $j$. But that is equivalent to $v_k v_j, v_i v_j \in A(D)$ and so to $v_k, v_i \in N^-(v_j)$ making $G$ a competition graph by definition.

Conversely, suppose $G = C(D)$, where $D$ is acyclic with $V(D) = \{v_1, \ldots, v_n\}$. By Lemma 4.5, we can assume the vertices of $D$ have been labeled so that $v_i v_j \in A(D)$ implies $i < j$. Check that $N^-(v_1) = Q_1, \ldots, N^-(v_n) = Q_n$ is an edge clique cover of $G$. Moreover, $v_i \in Q_j$ is equivalent to $v_i v_j \in A(D)$, which implies $i < j$.                    □

**Example 4.3 (continued)** For the competition graph $G$ in Figure 4.3, $Q_1 = \emptyset$, $Q_2 = \{v_1\}$, $Q_3 = \{v_1, v_2\}$, $Q_4 = \{v_1, v_3\}$, and $Q_5 = \{v_2, v_3, v_4\}$ form an edge clique cover as described in Theorem 4.6.

**Exercise 4.5** In the first paragraph of the proof of Theorem 4.6, show that $\mathcal{F}(\mathcal{E}) = \{S_1, \ldots, S_n\}$, the dual set representation of $G$ determined from $\mathcal{E}$, corresponds to the family of out-neighborhoods of $D$, with each $j \in S_i$ if and only if $v_j \in N_D^+(v_i)$.

In the second paragraph of the proof, $G$ has the set representation $\mathcal{F} = \{N^+(v_1), \ldots, N^+(v_n)\}$. Show that $\mathcal{E}(\mathcal{F}) = \{G_{v_1}, \ldots, G_{v_n}\}$, the dual edge cover of $G$ determined from $\mathcal{F}$, corresponds to the family of in-neighborhoods of $D$, with each $G_{v_i} = N_D^-(v_i)$.

Competition graphs of arbitrary (not necessarily acyclic) digraphs are characterized in [Dutton & Brigham, 1983] as follows.

**Theorem 4.7 (Dutton & Brigham)** *A graph $G$ is a competition graph of an arbitrary digraph if and only if $G$ has an edge clique cover $\mathcal{E}$ such that $|\mathcal{E}| = |V(G)|$.*

**Proof.** This can be proved by a minor modification of the proof of Theorem 4.6. □

**Exercise 4.6** Fill in the details in the proof of Theorem 4.7.

Recall from section 1.3 that $\theta(G)$ denotes the minimum cardinality of an edge clique cover of $G$. Theorem 4.7 can be rephrased as follows.

**Corollary 4.8 (Dutton & Brigham)** *A graph $G$ is a competition graph of an arbitrary digraph if and only if $\theta(G) \leq |V(G)|$.* □

As an elegant, but nontrivial, modification of this corollary, [Roberts & Steif, 1983] shows that a graph $G$ is a competition graph of a *loopless* digraph if and only if $\theta(G) \leq |V(G)|$ and $G \not\cong K_2$. The proof of Theorem 4.6 can be modified to produce the following more straightforward characterization.

**Exercise 4.7 (Dutton & Brigham)** Show that a graph $G$ is a competition graph of a loopless digraph if and only if $V(G)$ can be labeled as $\{v_1, \ldots, v_n\}$ and $G$ has an edge clique cover $\mathcal{E} = \{Q_1, \ldots, Q_n\}$ such that $v_i \in Q_j$ implies $i \neq j$.

[Fraughnaugh, Lundgren, Merz, Maybee, & Pullman, 1995] and [Guichard, 1998] describe competition graphs of strongly connected digraphs and of hamiltonian digraphs.

Given a competition graph $G$, the *competition number* of $G$, denoted $k(G)$, is the minimum number of isolated vertices that have to be added to $G$ to make it into a competition graph of an *acyclic* digraph.

**Exercise 4.8** (see [Roberts, 1978a]) Show that $k(G)$ is well defined by showing that at most $|E(G)|$ isolated vertices need to be added to any graph $G$ to make it into a competition graph of an acyclic digraph.

[Lundgren, 1989] and [Kim, 1993] survey the rich theory of competition numbers that has developed, and NP-completeness is shown in [Opsut, 1982]. This last paper also contains *Opsut's conjecture: If every $v \in V(G)$ has $N(v)$ coverable by at most two complete subgraphs of $G$, then $k(G) \leq 2$.* [Kim & Roberts, 1990, 1997] and [Wang, 1992, 1995b] further discuss competition numbers and variants of Opsut's conjecture.

Note that every competition graph of an acyclic digraph has at least one isolated vertex, since every acyclic digraph has at least one sink; the following result from [Roberts, 1978b] is a sort of converse to that.

**Exercise 4.9 (Roberts)** Suppose $G$ is a chordal graph that contains an isolated vertex. Prove that $G$ is a competition graph of an acyclic digraph. (Hint: Use the fact that $G$ must contain a simplicial vertex.)

The *common enemy graph* (or *resource graph*) of a digraph $D$ is the intersection graph $\Omega(\{N^-(v) : v \in V(D)\})$. *Competition-common enemy graphs*, in which $u, v \in V(D)$ are adjacent whenever *both $N^+(u) \cap N^+(v) \neq \emptyset$ and $N^-(u) \cap N^-(v) \neq \emptyset$*, and *niche graphs*, in which adjacency means that *either $N^+(u) \cap N^+(v) \neq \emptyset$ or $N^-(u) \cap N^-(v) \neq \emptyset$*, have been studied extensively, along with the corresponding analogues of competition numbers—again, see [Lundgren, 1989] and [Kim, 1993] for details and results, along with [Anderson, 1995], [Anderson, Jones, Lundgren, & Seager, 1991], [Hefner, Jones, Kim, Lundgren, & Roberts, 1991], and [Wang, 1995a]. [Raychaudhuri & Roberts, 1985] discusses other generalizations and applications of competition graphs.

## 4.3   Interval Competition Graphs

The impetus behind the intensive study of competition graphs was Joel Cohen's provocative 1968 observation, leading to the book [Cohen, 1978], that naturally occurring food webs tend to have interval competition graphs. Insofar as this is true, there would be potential ramifications for the ecological notion of "niche space." [Cohen & Palka, 1990] and [Cohen, Briand, & Newman, 1990] are recent sources describing the literature spawned by these questions and the status of Cohen's observation. (Building on the present section, we continue the food web story in the middle of section 6.2.)

The fundamental open problem in this area is to characterize those acyclic digraphs whose competition graph is an interval graph. [Lundgren, 1989] discusses this in detail, but we only include the one following result from [Lundgren & Maybee, 1984].

Given an acyclic digraph $D$, a *competition cover* $\mathcal{D} = \{D_1, \ldots, D_m\}$ of $D$ is a set of subsets of $V(D)$ such that, for all $v_i, v_j \in V(D)$, both $v_i, v_j \in D_\ell$ for some $D_\ell \in \mathcal{D}$ if and only if both $v_i v_k, v_j v_k \in A(D)$ for some $v_k \in V(D)$. A competition cover $\mathcal{D}$ has a *consecutive ranking* if its members can be arranged as vertices of a path $P$ such that, for each $v_i \in V(G)$, $\{D_j \in \mathcal{D} : v_i \in D_j\}$ induces a subpath of $P$.

**Theorem 4.9 (Lundgren & Maybee)** *An acyclic digraph has an interval competition graph if and only if it has a competition cover that has a consecutive ranking.*

**Proof.** Suppose $D$ is an acyclic digraph with competition graph $C(D)$.

First suppose $C(D)$ is an interval graph, say with clique path $P$ as in section 3.1. Take $\mathcal{D} = V(P)$. Then $v_i, v_j \in D_\ell \in \mathcal{D}$ for some $\ell$ if and only if $v_i v_j \in E(C(D))$, and so if and only if $v_i v_k, v_j v_k \in A(D)$ for some $k$. Moreover, $P$ determines a consecutive ranking of $\mathcal{D}$.

Conversely, suppose $\mathcal{D}$ is a competition cover of $D$ that has a consecutive ranking by a path $P$. Then $v_i, v_j \in D_\ell \in \mathcal{D}$ for some $\ell$ if and only if $v_i v_k, v_j v_k \in A(D)$ for some $k$, and so if and only if $v_i v_i \in E(C(D))$. So $\mathcal{D}$ is an edge clique cover of $C(D)$ and $P$ is a path representation for $C(D)$. Therefore, $C(D)$ is an interval graph. $\square$

**Exercise 4.10** Use Theorem 4.9 to show that the digraph in Figure 4.3 has an interval competition graph, but the digraph produced from the graph on the left in Figure 4.4 by directing each edge downward does not.

As far back as [Cohen, 1978], questions were also raised about digraphs that have *chordal* competition graphs. [Sugihara, 1984] considers various rationales to explain why competition graphs (and common enemy graphs) of naturally occurring food webs might tend to be chordal. See also [Pimm, 1991] for an ecology textbook's introduction to interval and chordal competition graphs and [McKee, 1995a] for several graph-related concepts that are potentially relevant.

[Lundgren & Merz, 1994] contains more characterizations of digraphs that have interval or chordal competition graphs. [Lundgren, Merz, & Rasmussen, 1993] contains characterizations of digraphs that have interval or chordal squared graphs. [Lundgren, Maybee, Merz, & Rasmussen, 1995] discusses digraphs that have interval or chordal two-step graphs.

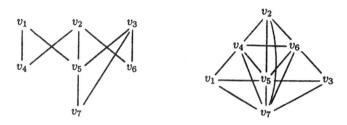

Figure 4.4: *A poset with its upper bound graph.*

## 4.4   Upper Bound Graphs

A *partially ordered set*, or *poset*, $(X, <)$ consists of a nonempty set $X$ with an irreflexive, transitive binary relation $<$ defined on it. The *upper bound graph of* $(X, <)$ is the graph $G$ with $V(G) = X$ and $uv \in E(G)$ if and only if $u \neq v$ and there exists $w \in X$ such that $u \leq w$ and $v \leq w$; in other words, $u$ and $v$ are distinct and have a common upper bound. A graph is an *upper bound graph* if it is isomorphic to the upper bound graph of some poset.

**Example 4.4** The figure on the left in Figure 4.4 shows a poset, where $v_i < v_j$ if and only if there is a downward path from $v_j$ to $v_i$. The upper bound graph of this poset is the graph on the right.

**Exercise 4.11** For any acyclic digraph $D$ and $v \in V(D)$, define the *ancestor set* of $v$ to be the subdigraph of $D$ induced by $\{w \neq v :$ there is a directed $w$-to-$v$ path in $D\}$. Show that a graph is an upper bound graph if and only if it is the intersection graph of ancestor sets of some acyclic digraph.

Upper bound graphs were introduced and characterized in [McMorris & Zaslavsky, 1982].

**Theorem 4.10 (McMorris & Zaslavsky)** *A graph $G$ is an upper bound graph if and only if $G$ has an edge clique cover $\mathcal{E} = \{Q_1, \ldots, Q_k\}$ such that, for each $j \in \{1, \ldots, k\}$, there exists $v_j \in V(G)$ such that $v_j \in Q_j$, but $v_j \notin Q_i$ for $i \neq j$.*

**Proof.** First suppose $G$ is the upper bound graph of $(\{v_1, \ldots, v_n\}, <)$. Without loss of generality, assume that $v_1, \ldots, v_k$ $(k \geq 1)$ are the maximal elements of $(X, <)$. For each $i \in \{1, \ldots, k\}$, let $Q_i = \{v_j : v_j \leq v_i\}$. It is easy to check that $Q_1, \ldots, Q_k$ form an edge clique cover of $G$ and each $v_j \in Q_j$. For $i \neq j$ in $\{1, \ldots, k\}$, $v_j$ maximal implies that $v_j \not< v_i$ and so $v_j \notin Q_i$.

Conversely, suppose $\mathcal{E}$ is an edge clique cover of $G$ as described in the theorem. Define $<$ on $V(G) = \{v_1, \ldots, v_n\}$ by $v_j < v_i$ if and only if $i \leq k < j$ and $v_j \in Q_i$. Then $v_i v_j \in E(G)$ with $i < j$ if and only if either $i \leq k < j$ with $v_i, v_j \in Q_i$ or there exists $h \leq k < i < j$ with $v_i, v_j \in Q_h$; these happen if and only if $v_i$ and $v_j$ have a common upper bound (respectively, $v_i$ or $v_h$). $\square$

Notice that, in the poset produced in the second part of the above proof, each $v_i$ is either a maximal element (when $i \leq k$) or a minimal element (when $i > k$); that is, "height-one posets suffice."

**Exercise 4.12** Find the edge clique cover $\mathcal{E}$ as in Theorem 4.10 for the upper bound graph on the right in Figure 4.4. Then find the poset $(X, <)$ for $G$ as produced in the proof. Also, show that removing vertex $v_2$ from $G$ would produce a graph that is not an upper bound graph.

**Exercise 4.13 (McMorris & Zaslavsky)** Show that the edge clique cover $\mathcal{E}$ in the statement of Theorem 4.10 can always be required to consist of maxcliques of $G$.

Recall from section 2.2 that a simplicial vertex is defined to be a vertex whose neighbors induce a complete subgraph (which may be the null subgraph).

**Exercise 4.14** (see [Bergstrand & Jones, 1988] and [Cheston, Hare, Hedetniemi, & Laskar, 1988]). Show that a graph is an upper bound graph if and only if every edge is in the closed neighborhood of a simplicial vertex.

The following more fundamental characterization of upper bound graphs appeared in [Lundgren & Maybee, 1983b].

**Theorem 4.11 (Lundgren & Maybee)** *A graph $G$ is an upper bound graph if and only if $V(G)$ can be labeled as $\{v_1, \ldots, v_n\}$ and $G$ has an edge clique cover $\mathcal{E} = \{Q_1, \ldots, Q_n\}$ such that both the following hold:*
*(1) each $v_i \in Q_i$;*
*(2) if $v_i \in Q_j$, then $j \leq i$ and $Q_i \subseteq Q_j$.*

**Proof.** This can be proved by modifying the proof of Theorem 4.10, using ideas from Exercise 4.11 and the proof of Theorem 4.6. $\square$

**Exercise 4.15** Prove Theorem 4.11.

[Scott, 1986] characterizes those posets that have interval and chordal upper bound graphs. [McMorris & Myers, 1983] discusses upper bound graphs that correspond to a unique poset.

[Lundgren, Maybee, & McMorris, 1988] contains various related topics concerning upper bound graphs (and similarly defined lower bound graphs) in relation to competition graphs (and common enemy graphs). [Bergstrand & Jones, 1989], [Bergstrand, Jones, & Sherman, to appear], and [Era & Tsuchiya, 1997, 1998] discuss more relations between upper and lower bound graphs.

# Chapter 5

# Threshold Graphs

Recall from section 2.5 that a graph $G$ is a split graph if $V(G)$ can be partitioned into $Q \cup I$, where $Q$ induces a complete graph and $I$ induces an edgeless graph (that is, $I$ is an independent set). Threshold graphs are special split graphs that were introduced in [Chvátal & Hammer, 1973] and that have been extensively studied since that time. In keeping with the style of the previous chapters, this chapter will provide only a short introduction to threshold graphs; [Mahadev & Peled, 1995] is a very nice comprehensive study.

## 5.1  Definitions and Characterizations

The definition that we give in this section is from [Chvátal & Hammer, 1973], in which set-packing problems are studied. For each vertex $v$ of a graph $G$, let $w_v$ denote a nonnegative real number, the *weight* of $v$. A graph $G$ is a *threshold graph* if there is an assignment of weights to the vertices of $G$ and a nonnegative real number $t$, the *threshold*, such that, for every $X \subseteq V(G)$, $X$ is an independent set if and only if $\sum_{v \in X} w_v \le t$—in other words, if weights can be assigned to the vertices of $G$ so that a subset of vertices is independent if and only if the total weight of the set is no greater than a certain constant threshold. Figure 5.1 shows two threshold graphs with weight assignments and thresholds.

The notion of degree partition of a vertex set is crucial to the understanding of threshold graphs. Let $G$ be a graph whose nonisolated vertices have the distinct degrees $\delta_1 < \delta_2 < \cdots < \delta_m$. Set $\delta_0 = 0$ and $\delta_{m+1} = |V|-1$, and let $D_i$ be the set of all vertices having degree $\delta_i$ for $i = 0, \ldots, m$. The sequence $D_0, \ldots, D_m$ is called the *degree partition* of $G$.

Figure 5.1: *Two threshold graphs having thresholds 4 and 7, respectively, using $v_{(i)}$ to denote that vertex $v$ has weight $w_v = i$.*

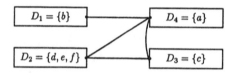

Figure 5.2: *Another view of the graph on the right in Figure 5.1, as explained in the text.*

**Example 5.1** The threshold graph on the left in Figure 5.1 has $m = 2$ with $\delta_1 = 1$, $\delta_2 = 4$, $D_0 = \emptyset$, $D_1 = \{a, b, d, e\}$, and $D_2 = \{c\}$. The threshold graph on the right has $m = 4$ with $D_0 = \emptyset$, $D_1 = \{b\}$, $D_2 = \{d, e, f\}$, $D_3 = \{c\}$, and $D_4 = \{a\}$.

Figure 5.2 shows another view of the graph on the right in Figure 5.1, with its vertices now grouped into "cells" corresponding to the degree partition. The $D_i$'s in the left column represent independent sets, the $D_i$'s in the right column represent complete subgraphs, and a line between cells $D_i$ and $D_j$ means that every vertex in $D_i$ is adjacent to every vertex in $D_j$.

Notice that the graph in Example 5.1 is a split graph, with the union of the cells on the left in Figure 5.2 forming the independent set $I$ and the union of those on the right inducing the complete subgraph $Q$. Also notice that the open neighborhoods of the vertices in the left column of cells are nested with respect to set inclusion in that the open neighborhood of every vertex in the left column is contained in the open neighborhood of every vertex below it; similarly, the closed neighborhoods of vertices in the right column are nested in that the closed neighborhood of every vertex in the right column is contained in the closed neighborhood of every vertex above it. One of the consequences of Theorem 5.1 will be that this sort of structure characterizes threshold graphs. This theorem is from [Chvátal & Hammer, 1973], with the equivalence of conditions (1) and (3) independently found in [Henderson & Zalcstein, 1977]).

**Theorem 5.1 (Chvátal & Hammer)** *Let $G = (V, E)$ be a graph with degree partition $D_0, \ldots, D_m$. Then the following statements are equivalent:*

(1) *$G$ is a threshold graph;*

(2) *for $x \in D_i$ and $y \in D_j$, $xy \in E$ if and only if $i + j > m$;*

(3) *there exist nonnegative integer weights $w_v$ and threshold $t$ such that, for distinct vertices $u$ and $v$, $uv \in E$ if and only if $w_u + w_v > t$;*

(4) *$G$ does not contain $P_4$, $C_4$, or $2K_2$ as an induced subgraph;*

(5) *$G$ is a split graph where the open neighborhoods of the vertices of the independent set $I$ can be nested with respect to set inclusion;*

(6) *$G$ can be obtained from $K_1$ by recursively adding either an isolated vertex or a vertex adjacent to every existing vertex.*

**Proof.** $(1 \Rightarrow 2)$: Assume $G$ is a threshold graph with weights $w_v$, threshold $t$, and degree partition $D_0, \ldots, D_m$, and suppose $0 < i \leq j \leq m$. Our proof is by induction on $m$, with the $m = 0$ case—when $G$ is edgeless—immediate. Now assume that $y \in D_m$ and $x \notin D_0$. Then there is a vertex $z$ such that $xz \in E(G)$, so that

$$t < w_x + w_z \leq w_x + w_y,$$

which implies that $xy \in E(G)$. Thus every vertex in $D_m$ is adjacent to every nonisolated vertex, and so $\delta_m = |V| - |D_0| - 1$. This shows condition (2) when $j = m$. Exercise 5.1 will show that $\delta_1 = |D_m|$, thus showing condition (2) when $i = 1$. Suppose $m > 1$, and let $V' = V - D_0 - D_m$ and $G'$ be the subgraph of $G$ induced by $V'$. Then $G'$ is a threshold graph with degree partition $D'_0, \ldots, D'_{m-2}$ where each $D'_i = D_{i+1}$. The induction hypothesis can then be used on $G'$ to show that condition (2) holds when $j = m - 1$ or $i = 2$; repetition shows that condition (2) holds in general.

$(2 \Rightarrow 3)$: This follows by assigning the weight $j$ to every vertex in $D_j$ and letting $t = m$.

$(3 \Rightarrow 4)$: Suppose each vertex $v \in V(G)$ is assigned weight $w_v$ and there is a threshold $t$ as in condition (3). Suppose $a, b, c, d \in V(G)$ with $ab, cd \in E(G)$ while $ad, bc \notin E(G)$. Then $w_a + w_b > t$, $w_a + w_d \leq t$, $w_c + w_d > t$, and $w_b + w_c \leq t$, an inconsistent set of inequalities.

$(4 \Rightarrow 5)$: Let $Q$ be a largest maxclique of $G$ and $I = G \backslash Q$. Suppose there exist $x, y \in I$ such that $xy \in E(G)$. Because $Q$ is a largest maxclique, there would exist vertices $u, v \in Q$ (possibly $u = v$) such that $xu, yv \notin E(G)$, leading in every case to an induced $P_4$, $C_4$, or $2K_2$. Thus $G$ is a split graph. Now let $x, y \in I$ have open neighborhoods $N(x)$ and $N(y)$. If $N(x) \not\subseteq N(y)$ and $N(y) \not\subseteq N(x)$, then it is easy to see that $x$ and $y$ would be in an induced $P_4$ or $2K_2$ in $G$.

$(5 \Rightarrow 6)$: Suppose $G$ satisfies condition (5). Since the removal of an isolated vertex or a vertex adjacent to every other vertex results in a graph that still satisfies condition (5), it suffices to show that $G$ contains such a vertex. Assume $I \neq \emptyset$ and $G$ has no isolated vertices. Pick $x \in I$ such that $N(x) \subseteq N(y)$ for every $y \in I$. Then any $z \in N(x)$ is adjacent to every vertex in $G$.

$(6 \Rightarrow 1)$: Assume $G$ satisfies condition (6) and proceed by induction on $|V(G)|$. Clearly $G$ is a threshold graph if $|V(G)| = 1, 2, 3$. Now assume that $G$ is a threshold graph with weights $w_v$ and threshold $t$. If we add a vertex $x$ that is adjacent to all the vertices of $G$, assign $x$ the weight $t$ and leave the other weights and threshold unchanged. If we add an isolated vertex $y$ to $G$, then assign weight $2w_v$ to each $v \in V(G)$, make the new threshold $2t + 1$, and assign to $y$ the weight 1.                                      $\square$

**Exercise 5.1** In the $(1 \Rightarrow 2)$ step of the proof of Theorem 5.1, show that vertices in $D_1$ are only adjacent to vertices in $D_m$.

Notice that condition (2) of Theorem 5.1 means that, for each $v \in D_k$,

$$N(v) = \bigcup_{j=1}^{k} D_{m+1-j} \text{ for } k = 1, \ldots, \lfloor m/2 \rfloor$$

and

$$N[v] = \bigcup_{j=1}^{k} D_{m+1-j} \text{ for } k = \lfloor m/2 \rfloor + 1, \ldots, m.$$

Therefore, the typical threshold graph $G$ has the structure shown in Figure 5.3, generalizing Figure 5.2: $D_0, \ldots, D_m$ is the degree partition of $G$ with $D_0$ possibly empty and $D_{\lceil m/2 \rceil}$ present only if $m$ is odd. A line between cells $D_i$ and $D_j$ means that every vertex in $D_i$ is adjacent to every vertex in $D_j$. The $D_i$'s in the left column represent independent sets, with the open neighborhoods of their vertices ordered by inclusion downward, and the $D_i$'s in the right column represent complete subgraphs, with the closed neighborhoods of their vertices ordered by inclusion upward.

**Exercise 5.2** Show that $G$ is a threshold graph if and only if $V(G)$ can be ordered $v_1, \ldots, v_n$ such that $v_i$ is adjacent to either none or all of the vertices in the subgraph induced by $\{v_i, \ldots, v_n\}$.

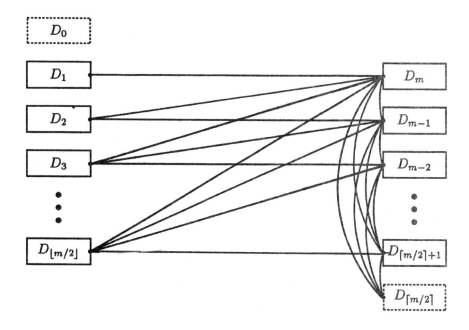

Figure 5.3: *The structure of a typical threshold graph.*

Also, find such an ordering of the vertices of the graph on the right in Figure 5.1.

**Corollary 5.2** *The complement of a threshold graph is a threshold graph.*

**Proof.** This follows immediately from condition (4) of Theorem 5.1 and the facts that $P_4$ is self-complementary and that $C_4$ and $2K_2$ are complements of each other. □

**Exercise 5.3** Show that a graph is a threshold graph if and only if neither it nor its complement contains $P_4$ or $C_4$ as an induced subgraph. (Such $P_4, C_4$-free graphs are discussed further in section 7.9.)

## 5.2 Threshold Graphs as Intersection Graphs

Since Theorem 5.1 shows that threshold graphs are split graphs, the intersection characterization of split graphs in Corollary 2.17 can be used to produce the following characterization of threshold graphs.

**Exercise 5.4** Show that a graph is a threshold graph if and only if it is the intersection graph of a set of distinct substars of a star where the substars containing the center of the star are nested with respect to set inclusion.

Let $G$ be an interval graph. A *threshold interval representation* for $G$ is an interval representation for $G$ that consists of a family of intervals $\{J_{v_i}\}$ such that $J_{v_i}$ is either the interval $[0, r_i]$ or the trivial interval $[s_i, s_i]$ where $s_j \neq s_k$ for all $j \neq k$ and where $s_j \neq r_k$ for all $j$ and $k$. The following is from [Mahadev & Peled, 1995].

**Theorem 5.3** *A graph is a threshold graph if and only if it is an interval graph with a threshold interval representation.*

**Proof.** Assume $G$ is an interval graph with a threshold interval representation. Then the nontrivial intervals in the representation correspond to vertices that induce a maxclique $Q$ of $G$, the trivial intervals correspond to vertices that form an independent set $I$ in $G$, and the neighborhoods of vertices in $I$ are nested in the order in which their representing trivial intervals appear along the real line. Therefore, $G$ is a threshold graph by part (5) of Theorem 5.1.

Conversely, suppose that $G$ is a threshold graph. By part (5) of Theorem 5.1, $G$ is a split graph with $V(G)$ partitioned into the complete subgraph $Q$ and the independent set $I$ with the neighborhoods of $I$ nested. Assign each vertex $v$ in $Q$ an interval $J_v = [0, r_v]$ such that $N[u] \subseteq N[v]$ if and only if $r_u \leq r_v$ for all $u, v \in Q$. For each $w \in I$, let $r(w) = \max\{r_v : wv \in E(G)\}$. Each such $w$ can be assigned a trivial interval $J_w$ that is a small distance $\epsilon_w$ to the left of $r(w)$ in such a way that the intervals form a threshold interval representation for $G$.                                                                          $\square$

**Example 5.2** To illustrate the proof of Theorem 5.3, the threshold graph on the left in Figure 5.1 could receive the threshold interval representation determined by $0 < s_a < s_b < s_d < s_e < r_c$. The threshold graph on the right could receive the threshold interval representation determined by $0 < s_d < s_e < s_f < r_c < s_b < r_a$.

**Exercise 5.5** Verify that if each of the cells in the general threshold graph shown in Figure 5.3 is the singleton $D_i = \{v_i\}$ (and if $m$ is odd), then any choice of $r_i$'s and $s_i$'s as in Figure 5.4 would determine a threshold interval representation of the graph.

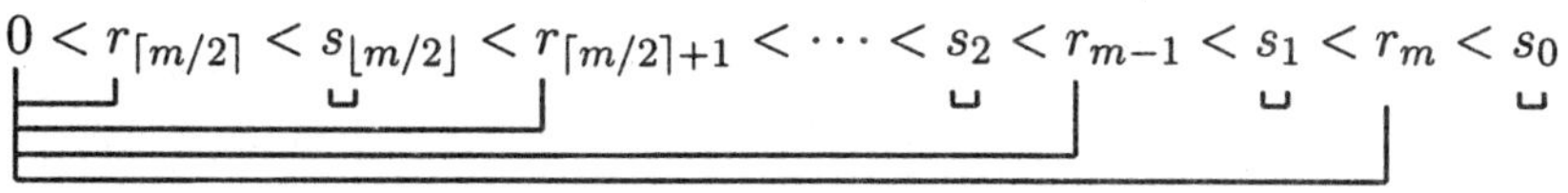

$$0 < r_{\lceil m/2 \rceil} < s_{\lfloor m/2 \rfloor} < r_{\lceil m/2 \rceil+1} < \cdots < s_2 < r_{m-1} < s_1 < r_m < s_0$$

Figure 5.4: *A threshold interval representation for a typical threshold graph as in Figure 5.3.*

Figure 5.5: *An interval graph that requires only two lengths of intervals, yet is not a threshold graph.*

Recall that a unit interval graph is an interval graph with an interval representation using intervals all of the same length. There are threshold graphs that are unit interval graphs ($K_n$, for example) and others that are not unit interval graphs ($K_{1,3}$, for example). However, as first observed in [Leibowitz, 1978], a threshold graph will never require more than two distinct lengths of intervals in its interval representation.

**Theorem 5.4 (Leibowitz)** *Every threshold graph has an interval representation whose intervals have at most two distinct lengths.*

**Proof.** Let $G$ be a threshold graph with $V(G)$ partitioned into the maxclique $Q$ and independent set $I$, and suppose $\{J_v\}$ is a threshold interval representation as constructed in the proof of Theorem 5.3. By that construction, there exists a $z \in Q$ such that $r_x \leq r_z$ for all $x \in Q$. For each $u \in Q$ assign the interval $J'_u = [r_u - r_z, r_u]$, while for each $w \in I$ let $J'_w = J_w$. Then $J'_v$ is an interval representation for $G$ using only the two lengths $r_z$ and 0. $\square$

The converse to Theorem 5.4 fails since the nonthreshold graph $P_4$ is a unit interval graph.

**Exercise 5.6** Show that the nonthreshold graph in Figure 5.5 is an interval graph whose interval representations require two but only two different interval lengths. ([Skrien, 1984] characterizes all such graphs.)

Figure 5.6: $P_4$ is a difference graph with weights $w_{x_1} = 3$, $w_{x_2} = 2$, $w_{y_1} = -2$, $w_{y_2} = -1$, and $T = 4$; $G_X$ is as in Theorem 5.5 using $X = \{x_1, x_2\}$.

## 5.3   Difference Graphs and Ferrers Digraphs

This section is a brief introduction to two types of graphs that are closely related to threshold graphs. [Mahadev & Peled, 1995] contains a more thorough treatment and extensive references.

Define a *difference graph* to be a graph $G = (V, E)$ such that each vertex $v \in V$ can be assigned a real number weight $w_v$ and there exists a positive real number $T$ such that both the following hold:

(1) $|w_v| < T$ for all $v \in V$;

(2) if $u \neq v$, then $uv \in E$ if and only if $|w_u - w_v| \geq T$.

Difference graphs were first introduced in [Hammer, Peled, & Sun, 1990], emphasizing their similarity to threshold graphs. Notice that every difference graph $G$ is bipartite, since $V(G)$ can be partitioned into the two independent sets $X = \{v : w_v \geq 0\}$ and $Y = \{u : w_u < 0\}$. Thus $K_3$ is a threshold graph that is not a difference graph. Figure 5.6 assigns weights to $P_4$ to show that it is a difference graph that, by Theorem 5.1, is not a threshold graph.

Let $G = (V, E)$ be any bipartite graph with $V$ partitioned into independent sets $X$ and $Y$, and define the split graph $G_X = (V, E \cup E_X)$ to have $E_X = \{uv : u, v \in X \text{ and } u \neq v\}$; see Figure 5.6.

**Theorem 5.5 (Hammer, Peled, & Sun)** *A graph $G = (V, E)$ is a difference graph if and only if there is a partition of $V$ into independent sets $X$ and $Y$ such that $G_X$ is a threshold graph.*

**Proof.** Let $G = (V, E)$ be a difference graph with vertex weights $w_v$ for $v \in V$ and with $V$ partitioned into the independent sets $X = \{v : w_v \geq 0\}$ and $Y = \{u : w_u < 0\}$. To each $v \in X$ assign the new weight $w'_v = T + w_v$, and to each $u \in Y$ assign the new weight $w'_u = -w_u$. Let $t = 2T$. If $x, v \in X$, then $w'_x + w'_v \geq t$ since both $w_x, w_v \geq 0$. If $x \in X$ and $y \in Y$, then $w'_x + w'_y \geq t$ if and only if $|w_x - w_y| \geq T$. If $y, u \in Y$, then $w'_y + w'_u < t$

Figure 5.7: *Forbidden configurations in a Ferrers digraph: in each, if the two solid arcs occur, then at least one of the two dashed arcs must occur.*

since both $|w_y|, |w_u| < T$. Therefore, condition (3) of Theorem 5.1 shows that $G_X$ is a threshold graph with weights $w'_v$ and threshold $t$.

For the converse, suppose $V$ is partitioned into independent sets $X$ and $Y$ such that $G_X$ is a threshold graph with degree partition $D_0, \ldots, D_m$; furthermore, we can assume that $X$ and $Y$ are chosen so that

$$Y = D_0 \cup \cdots \cup D_{\lfloor m/2 \rfloor} \quad \text{and} \quad X = D_{\lfloor m/2 \rfloor + 1} \cup \cdots \cup D_m.$$

To each $x \in D_i \subseteq X$ assign $w_x = i - \lceil m/2 \rceil$, and to each $y \in D_j \subseteq Y$ assign $w_y = j - \lfloor m/2 \rfloor - 1$. Thus for $u \neq v$, $uv \in E \cup E_X$ if and only if $w_u + w_v \geq 0$, and in addition $w_x \geq 0$ for $x \in X$ and $w_y < 0$ for $y \in Y$. Pick $T$ such that $T > \max\{w_x : x \in X\}$ and $T > \max\{-w_y : y \in Y\}$. To each $x \in X$ assign $w'_x = w_x$, and to each $y \in Y$ assign $w'_y = -w_y - T$. It is easy to check that conditions (1) and (2) of the definition of a difference graph are satisfied. □

**Theorem 5.6 (Hammer, Peled, & Sun)** *A graph is a difference graph if and only if it does not contain $K_3$, $C_5$, or $2K_2$ as an induced subgraph.*

**Exercise 5.7** Prove Theorem 5.6.

Introduced in [Riguet, 1951], a *Ferrers digraph* is a digraph $D = (V, A)$— with loops allowed—such that for all $w, x, y, z \in V$ (not necessarily distinct except that $w \neq y$ and $x \neq z$), it is *not* the case that $wx, yz \in A$ and $wz, yx \notin A$. (This says that $D$ satisfies Axiom 2 from section 3.4.2; a loopless digraph is a Ferrers digraph if and only if it is an interval order.) Figure 5.7 illustrates what is forbidden in Ferrers digraphs, taking the allowed coalescence of $w, x, y, z$ into account.

**Exercise 5.8** Show that the digraph on the left in Figure 3.4 is not a Ferrers digraph.

**Exercise 5.9** Show that $D = (V, A)$ is a Ferrers digraph if and only if the out-neighborhoods $N^+(v)$ for $v \in V$ are nested (or, equivalently, the in-neighborhoods $N^-(v)$ for $v \in V$ are nested).

**Exercise 5.10  (Cogis)** Show that if $D = (V, A)$ is a symmetric Ferrers digraph—"symmetric" meaning that $uv \in A$ if and only if $vu \in A$—then the underlying graph of $D$ (discarding any loops) is a threshold graph. ([Cogis, 1982] also shows how to go from an arbitrary threshold graph to a symmetric Ferrers graph with properly chosen loops.)

For a digraph $D = (V, A)$ with $V = \{v_1, \ldots, v_n\}$, define the *bipartite representation* of $D$ to be the bipartite graph $B(D)$ on $V(B(D)) = \{x_1, \ldots, x_n; y_1, \ldots, y_n\}$ with $x_i y_j \in E(B(D))$ exactly when $v_i v_j \in A$. For instance, the graph $G$ in Figure 7.6 is the bipartite representation of the digraph $D$ shown there.

**Exercise 5.11** Show that a digraph $D$ is a Ferrers digraph if and only if its bipartite representation $B(D)$ is a difference graph.

## 5.4   Some Applications of Threshold Graphs

The study of threshold graphs began in [Chvátal & Hammer, 1973, 1977] with applications to the "aggregation" of linear inequalities in integer programming and set packing problems. There have been several other application areas such as the synchronization of parallel processes in [Golumbic, 1978c], [Henderson & Zalcstein, 1977], and [Ordman, 1989], and to cyclic scheduling in [Koop, 1986]. This section focuses on an application in the social sciences that fits in more naturally with our approach to threshold graphs.

Suppose $S$ is a set of "subjects" and $I$ is a set of "items," where for example the $(S, I)$ pairs might be (students, tests), (soldiers, combat situations) or (people, opinion poll questions). Assume further that there is a binary relation $\rho$ between $S$ and $I$; for the three previous examples $\rho$ might be, respectively, "can pass," "fears," and "agrees with." It is often desired to linear order $S \cup I$ so that the $\rho$ relation is preserved. Formally, a *Guttman scale* is a mapping $g : S \cup I \mapsto \mathbb{R}$ such that for each $u \in S$ and $v \in I$, $u \rho v$ if and only if $g(u) < g(v)$.

Illustrating this with the polling example, the existence of a Guttman scale means that the people and opinions can be linearly ordered ("scaled") so that each person agrees with the opinions succeeding her in the scale and disagrees with each opinion preceding her in the scale.

We want to determine for which $S, I$, and $\rho$ a Guttman scale exists. To do this, consider the bipartite graph $G$ in which $V(G) = S \cup I$, $S$ and $I$ are independent and, for $u \in S$ and $v \in I$, $uv \in E(G)$ if and only if $u\rho v$. Let $G_S$ be the split graph formed from $G$ by adding all edges between vertices in $S$ in order to make $S$ complete.

The following theorem from [Cozzens & Leibowitz, 1984] and [Leibowitz, 1978] combines with Theorem 5.5 to show that the bipartite graph of a Guttman scale is a difference graph.

**Theorem 5.7 (Cozzens and Leibowitz)** *A Guttman scale exists if and only if $G_S$ is a threshold graph.*

**Proof.** Suppose $G_S$ is a threshold graph. Then by condition (6) of Theorem 5.1, the vertices of $G_S$ can be linearly ordered such that every vertex in $S$ is adjacent to every vertex preceding it in the order but to no others. The reverse of this order leads to a mapping $g$ that satisfies the definition of a Guttman scale.

Now assume that $G_S$ has a Guttman scale $g$ yet is not a threshold graph. By condition (4) of Theorem 5.1 and $G_S$ being split, there exists an induced $P_4$, say $a, b, c, d$ with $b, c \in S$ and $a, d \in I$. This implies the contradictory inequalities $g(d) \leq g(b) < g(a)$ and $g(a) \leq g(c) < g(d)$.                    □

[Cozzens & Leibowitz, 1987] contains further discussion of the connection between Guttman scales and graphs.

# Chapter 6

# Other Kinds of Intersection

Given a family $\mathcal{F} = \{S_1, \ldots, S_n\}$ of nonempty sets, the existence of edges in the intersection graph $\Omega(\mathcal{F})$ depends only on whether $S_i \cap S_j \neq \emptyset$. Clearly, however, the cardinality of the intersections and other features can also be relevant. This chapter considers three of the other kinds of intersection that have been worked on and shows for each how the basic theory from Chapter 1 can be modified and what analogues exist for various sorts of intersection graphs we have studied.

Chapter 7 contains other examples of different kinds of intersection.

## 6.1  $p$-Intersection Graphs

For each integer $p \geq 1$, the *p-intersection graph* $\Omega_p(\mathcal{F})$ of the family $\mathcal{F} = \{S_1, \ldots, S_n\}$ of subsets of a *finite* set $S$ is defined to be the graph $G$ having $V(G) = \mathcal{F}$ with $S_i S_j \in E(G)$ if and only if $i \neq j$ and $|S_i \cap S_j| \geq p$. A graph $G$ is a *p-intersection graph* if there exists a family $\mathcal{F}$ such that $G \cong \Omega_p(\mathcal{F})$, and $\mathcal{F}$ is then called a *p-intersection set representation* for $G$. Thus the 1-intersection graphs are precisely the ordinary intersection graphs on finite sets. The concept of the $p$-intersection graph was introduced in [Jacobson, McMorris, & Scheinerman, 1991]; see also [McKee, 1991a] and [Kim, McKee, McMorris, & Roberts, 1995].

**Example 6.1** Suppose $\mathcal{F} = \{S_1, \ldots, S_5\}$ where $S_1 = \{a, b, c\}$, $S_2 = \{b, c, d\}$, $S_3 = \{b, c, d, f, g\}$, $S_4 = \{c, d, e, f, g\}$, and $S_5 = \{a, d, e\}$. Then the 2-intersection graph $G \cong \Omega_2(\mathcal{F})$ is shown on the left in Figure 6.1, with a set-labeled version on the right. The graph $\Omega_3(\mathcal{F})$ corresponds to the path $S_2 S_3 S_4$ and two isolated vertices, $\Omega_4(\mathcal{F})$ corresponds to the edge $S_3 S_4$ and three isolated vertices, $\Omega_k(\mathcal{F})$ is edgeless for $k > 4$, and $\Omega_1(\mathcal{F})$ is complete.

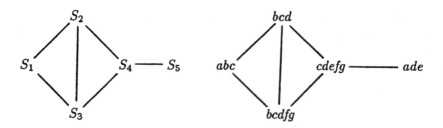

Figure 6.1: *A 2-intersection graph.*

**Exercise 6.1** Show that every graph is a $p$-intersection graph for every $p \geq 1$.

The key concept for $p$-intersection graph theory is a *$p$-edge clique cover* of a graph $G$, which is a family $\{V_1, \ldots, V_m\}$ of *not necessarily distinct* subsets of $V(G)$ such that, for every set $\{i_1, \ldots, i_p\}$ of $p$ *distinct* subscripts, $T = V_{i_1} \cap \cdots \cap V_{i_p}$ induces a complete subgraph of $G$—recall that $T$ may be the null subgraph of $G$—and such that the collection of sets of the form $T$ is an edge clique cover of $G$. The proof of Theorem 6.1 will show that $\{V_1, \ldots, V_m\}$ is a $p$-edge clique cover of $G$ if and only if $v_i v_j \in E(G)$ is equivalent to $v_i$ and $v_j$ being in at least $p$ common sets $V_i$. This shows that $p$-edge clique covers are what are called *$p$-generators* in [Chung & West, 1994].

**Exercise 6.2** Show that $V_1 = \{S_1, S_5\}$, $V_2 = \{S_1, S_2, S_3\}$, $V_3 = \{S_1, S_2, S_3, S_4\}$, $V_4 = \{S_2, S_3, S_4, S_5\}$, $V_5 = \{S_4, S_5\}$, $V_6 = \{S_3, S_4\}$, and $V_7 = \{S_3, S_4\}$ is a 2-edge clique cover of the graph $G$ in Example 6.1. Show that $V_1 = \{S_1, S_2\}$, $V_2 = \{S_1, S_3\}$, $V_3 = \{S_1, S_2, S_3\}$, $V_4 = \{S_2, S_3, S_4\}$, $V_5 = \{S_2, S_4\}$, $V_6 = \{S_3, S_4\}$, $V_7 = \{S_4, S_5\}$, and $V_8 = \{S_4, S_5\}$ is another 2-edge clique cover.

**Exercise 6.3** Check that being a 1-edge clique cover is the same as being an edge clique cover.

**Theorem 6.1** *Suppose $G$ is the $p$-intersection graph of $\mathcal{F} = \{S_1, \ldots, S_n\}$ on the set $S$. For each element $x \in S$, put $V_x = \{S_i \in V(G) : x \in S_i\}$. Then the family of these $V_x$'s forms a $p$-edge clique cover for $G$.*

**Proof.** Suppose $G$ and $\mathcal{F}$ are as in the theorem and $\{x_1, \ldots, x_p\}$ is a set of distinct elements of $S$ and $S_i, S_j \in T = V_{x_1} \cap \cdots \cap V_{x_p}$. Then

$\{x_1, \ldots, x_p\} \subseteq S_i \cap S_j$, so $|S_i \cap S_j| \geq p$, and so $S_i S_j \in E(G)$. Thus $T$ induces a complete subgraph (possibly the null subgraph) of $G$. To show the sets of form $T$ cover $E(G)$, suppose $S_i S_j \in E(G)$. This implies $|S_i \cap S_j| \geq p$ and thus there exist distinct $x_1, \ldots, x_p \in S_i \cap S_j$. By definition of the $V_x$'s, we have $S_i, S_j \in V_{x_1} \cap \cdots V_{x_p}$. $\qquad\square$

**Exercise 6.4** Show that the first 2-edge clique cover listed in Exercise 6.2 is the sort described in Theorem 6.1. Also check that we really have shown that $p$-edge clique covers are the same as $p$-generators (as defined above).

**Exercise 6.5** Suppose $G$ has $p$-edge clique cover $\{V_1, \ldots, V_m\}$. For each $v_i \in V(G) = \{v_1, \ldots, v_n\}$, define $R_i = \{j : v_i \in V_j\}$. Show that $R_1, \ldots, R_n$ is a $p$-intersection set representation for $G$.

For any graph $G$, the *p-intersection number* of $G$ is the minimum cardinality of a set $S$ such that $G$ is a $p$-intersection graph on $S$. The following is analogous to Theorem 1.6.

**Theorem 6.2** *For every graph $G$, the $p$-intersection number of $G$ equals the minimum cardinality of a $p$-edge clique cover of $G$.*

**Proof.** This follows by a similar argument to that used for Theorem 1.6, using Theorem 6.1 and Exercise 6.5. $\qquad\square$

Actually finding $p$-intersection numbers is hard, even in the $p = 2$ case; see [Chung & West, 1994], [Ganter, Gronau, & Mullin, 1994], [Jacobson, Kézdy, & West, 1995], and [Eaton, 1997]. For instance, Jacobson, Kézdy, and West show that the 2-intersection number of the $n$-vertex path $P_n$ is asymptotic to $2\sqrt{n}$. Also see [Brigham, Dutton, & McMorris, 1992, 1993], [Eaton, Gould, & Rödl, 1996], and [Füredi, 1997].

Paralleling section 1.3, define a graph to be a *p-clique graph* if it is isomorphic to the $p$-intersection graph of all maxcliques of some graph. It is not hard to show that the direct analogue of Theorem 1.12 holds: A graph is a $p$-clique graph if and only if it has a $p$-edge clique cover that satisfies the Helly condition; see [McKee, 1991a], which also shows that $p$-clique graphs are clique graphs.

**Exercise 6.6** Show that, for every $p \geq 1$, a graph is the $p$-intersection graph of a family of subtrees of a tree if and only if the graph is chordal.

Thus chordal graphs do not have interesting $p$-analogues, and the same is true for interval graphs, proper interval graphs, unit interval graphs, and line graphs. However, parts of intersection graph theory can have interesting $p$-intersection graph analogues in unexpected ways. In particular, section 4.2 motivates the most-studied topic in $p$-intersection graph theory: $p$-competition graphs, meaning graphs isomorphic to the $p$-intersection graph of the out-neighborhoods of vertices of a digraph. [Kim, McKee, McMorris, & Roberts, 1995] is the most general source, although not the earliest. For instance, the following direct analogue of Theorem 4.7, the Dutton and Brigham characterization, appeared in [Isaak, Kim, McKee, McMorris, & Roberts, 1992].

**Theorem 6.3** *A graph $G$ is the p-competition graph of an arbitrary digraph if and only if $G$ has a p-edge clique cover of cardinality $|V(G)|$.*

**Proof.** First suppose $G$ is the $p$-competition graph of $D$, where $V(G) = V(D) = \{v_1, \ldots, v_n\}$. It is easy to check that $\{N^-(v_i) : 1 \leq i \leq n\}$, the family of all in-neighborhoods of $D$, is a $p$-edge clique cover of $G$.

Conversely, suppose $G$ has a $p$-edge clique cover $\{V_1, \ldots, V_r\}$, where $r \leq n$; since repetitions are allowed, we can assume that $r = n$. Define a digraph $D$ with $V(D) = V(G)$ with $v_i v_j \in A(D)$ if and only if $v_i \in V_j$. It is easy to check that $G$ is the $p$-competition graph of $D$.          $\square$

**Exercise 6.7** Show that $C_4$ is not the 2-competition graph of an arbitrary digraph.

However, not everything goes over directly from competition graphs to $p$-competition graphs. For instance, Theorem 4.7 made it easy to tell which complete bipartite graphs $K_{m,n}$ are competition graphs of arbitrary digraphs: precisely those for which $mn \leq m+n$. But only partial results are known even for which complete bipartite graphs are 2-competition graphs of arbitrary digraphs; for instance [Isaak, Kim, McKee, McMorris, & Roberts, 1992] shows that $K_{2,n}$ is the 2-competition graph of an arbitrary digraph if and only if $n = 1$ or $n \geq 9$, and [Jacobson, 1992] shows that $K_{m,m}$ is the 2-competition graph of an arbitrary digraph if and only if $m = 1$. There are also analogous questions for $p$-competition numbers of graphs in [Kim, McKee, McMorris, & Roberts, 1993].

[Kim, McKee, McMorris, & Roberts, 1995] also shows the direct analogues—replacing "competition" with "$p$-competition" and "edge clique cover" with "$p$-edge complete cover"—of Theorem 4.6, for acyclic digraphs, and Exercise 4.7, for loopless digraphs. But no simple analogue is known for the characterization in [Roberts & Steif, 1983] that a graph $G$ is a competition graph of a *loopless* digraph if and only if $\theta(G) \leq |V(G)|$ and $G \not\cong K_2$.

See [Major & McMorris, 1990] for $p$-intersection graph analogues of various other intersection graph concepts. [Lundgren, McKenna, Merz, & Rasmussen, 1995, to appear], [Lundgren, McKenna, Langley, Merz, & Rasmussen, 1997], and [Anderson, Langley, Lundgren, McKenna, & Merz, 1994] are some of the recent papers on $p$-competition graphs of special types of digraphs and other related topics.

[Eaton & Grable, 1996] and [McMorris & Wang, 1996] discuss notions related to $p$-intersection graphs, such as having $S_i S_j \in E(G)$ depend on the congruence class of $|S_i \cap S_j|$ modulo a given number—in particular, when $|S_i \cap S_j|$ is odd.

## 6.2 Intersection Multigraphs and Pseudographs

The *intersection multigraph* $\Omega_\mu(\mathcal{F})$ of the family $\mathcal{F} = \{S_1, \ldots, S_n\}$ of subsets of a finite set $S$ is the multigraph $M$ having $V(M) = \mathcal{F}$ with $S_i$ and $S_j$ joined by $|S_i \cap S_j|$ parallel edges whenever $i \neq j$. When $|S_i \cap S_j| \geq 1$, $S_i S_j$ is a *multiple edge* with *multiplicity* $|S_i \cap S_j|$. Let $E(M)$ be the set of all multiple edges of $M$. A multigraph $M$ is an *intersection multigraph* if there exists a family $\mathcal{F}$ such that $M \cong \Omega_\mu(\mathcal{F})$.

For any multigraph, its *underlying graph* is obtained by replacing each multiple edge by a *simple edge*, an edge of multiplicity one. Thus the ordinary intersection graph $\Omega(\mathcal{F})$ is the underlying graph of the intersection multigraph $\Omega_\mu(\mathcal{F})$.

**Example 6.2** Suppose $\mathcal{F} = \{S_1, \ldots, S_5\}$ where $S_1 = \{a, b, c\}$, $S_2 = \{b, c, d\}$, $S_3 = \{b, c, d, f, g\}$, $S_4 = \{c, d, e, f, g\}$, and $S_5 = \{a, d, e\}$. Then the intersection multigraph $M \cong \Omega_\mu(\mathcal{F})$ is shown on the left in Figure 6.2, with a set-labeled version on the right. Notice how this multigraph simultaneously displays all the $\Omega_i(\mathcal{F})$'s from Example 6.1.

**Exercise 6.8** Show that every multigraph is an intersection multigraph.

An *edge clique partition* of a multigraph $M$ is a family $\{Q_1, \ldots, Q_n\}$ of *not necessarily distinct* complete subgraphs of the underlying graph of $M$ such that each $v_i v_j \in E(M)$ has multiplicity $|\{k : v_i v_j \in E(Q_k)\}|$. Thus $M$ can be thought of as the superposition of $Q_1, \ldots, Q_n$, identifying vertices while collecting edges into bundles of parallel edges.

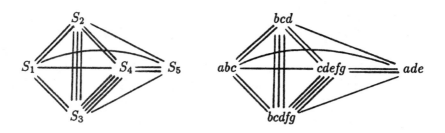

Figure 6.2: *An intersection multigraph.*

**Example 6.2 (continued)** The complete subgraphs induced by $\{S_1,$ $S_2, S_3,$ $S_4,$ $S_5\}$, $\{S_1, S_2, S_3\}$, $\{S_2, S_3, S_4\}$, $\{S_4, S_5\}$, $\{S_3, S_4\}$, and $\{S_3, S_4\}$ form one edge clique partition of the multigraph $M$ in Figure 6.2. Another consists of the 19 $K_2$'s.

Paralleling section 1.4, define a multigraph to be the *clique multigraph of a graph* $G$ if it is isomorphic to the intersection multigraph of all max-cliques of $G$. It is straightforward to modify the proof of Theorem 1.12 from [Roberts & Spencer, 1971] to show that a multigraph is a clique multigraph of a graph if and only if it has an edge clique partition that satisfies the Helly condition. [McKee, 1991c] contains more about clique multigraphs.

**Exercise 6.9** Use the above-mentioned characterization of clique multigraphs to show that removing an edge from $K_4$ produces a clique graph that is not a clique multigraph.

The analogues of Theorem 4.2, characterizing squared graphs, and Theorem 4.4, characterizing two-step graphs, are also straightforward; see [McKee, 1990a]. In fact, [Harary & McKee, 1994] shows that the "squared multigraph" of a chordal graph is particularly nice in that its "square root"—the chordal graph whose square is the given multigraph—can be uniquely constructed. [Prisner, to appear] shows an advantage of considering "triangle multigraphs"—intersection multigraphs of the $K_3$'s of a graph—rather than "triangle graphs." [McKee, 1989] introduces "upper bound multigraphs," showing how they determine their associated posets up to isomorphism. [Anderson, Jones, Lundgren, & McKee, 1990] discusses "competition multigraphs" and "multicompetition numbers," and [Bylka & Komar, 1997] considers intersection numbers of intersection multigraphs.

[McKee, 1991b] characterizes *chordal multigraphs*, the intersection multigraphs of subtrees of a tree, and *interval multigraphs*, the intersection multi-

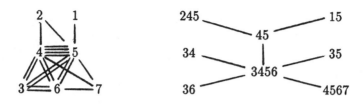

Figure 6.3: *A chordal multigraph with a tree representation.*

graphs of subpaths of a path. For instance, Figure 6.3 shows a chordal multigraph with a tree representation.

We state and prove the characterization of chordal multigraphs; the corresponding theorem for interval multigraphs is more involved.

**Theorem 6.4 (McKee)** *A multigraph $M$ with underlying graph $G$ is a chordal multigraph if and only if both the following hold:*

*(1) the multiplicity of each $uv \in E(M)$ is greater than or equal to the number of maxcliques in $G$ that contain both $u$ and $v$;*

*(2) $G$ is chordal.*

**Proof.** First suppose that $M$ is the intersection multigraph of a family of subtrees of some tree $T$ and $G$ is the underlying graph of $M$. Since $T$ is also a tree representation for $G$, condition (2) follows from Theorem 2.4. If $uv \in E(M)$ has multiplicity $\mu$, then $|T_u \cap T_v| = \mu$ and $\{u, v\}$ will be in $\mu$ of the vertices of $T$ that are maxcliques of $G$ (by the proof of Theorem 2.1). Condition (1) then follows.

Conversely, suppose $M$ has underlying graph $G$ and satisfies conditions (1) and (2). Construct a multigraph $M^+$ from $M$ as follows: For *each* edge $uv \in E(M)$ with multiplicity $\mu$, let $k$ be the number of maxcliques of $G$ that contain $uv$; by condition (1), we can create $\mu - k \geq 0$ new simplicial vertices, each joined (only) to $u$ and $v$ by simple edges in $M^+$. (Thus $M^+$ satisfies condition (1) with equality always holding.) Let $G^+$ be the underlying graph of $M^+$, and note that $G^+$ is chordal by condition (2). Let $T^+$ be a clique tree for $G^+$, and construct $T$ from $T^+$ by removing occurrences of vertices in $V(M^+) \setminus V(M)$ from vertices of $T^+$. Then it is straightforward to verify that $M \cong \Omega_\mu(\{T_v : v \in V(M)\})$.    $\square$

**Exercise 6.10** Show that in a chordal multigraph $M$, every circuit of length greater than or equal to four must contain at least two (possibly parallel) chords.

**Exercise 6.11** Show that the multigraph obtained by removing edge $S_1S_5$ from the multigraph in Figure 6.2 is not a chordal multigraph. Also show that the tree representation for the chordal multigraph in Figure 6.3 can be constructed as in the proof of Theorem 6.4.

We now return to the topic of section 4.3, the premise of [Cohen, 1978] that "...it is possible for niche overlaps to be described in a one-dimensional niche space if and only if the niche overlap graph [competition graph] is an interval graph." Define a *competition multigraph* of a digraph $D$ to be a multigraph isomorphic to the intersection multigraph of the out-neighborhoods of vertices of $D$, and recall that a food web is an *acyclic* digraph.

**Theorem 6.5** *A multigraph $M$ is a competition multigraph of an acyclic digraph if and only if $V(M)$ can be labeled as $\{v_1, \ldots, v_n\}$ and $M$ has an edge clique partition $\mathcal{E} = \{Q_1, \ldots, Q_n\}$ such that $v_i \in Q_j$ implies $i < j$.*

**Proof.** This follows by a modification of the proof of Theorem 4.6. $\square$

**Exercise 6.12** Complete the details of the proof of Theorem 6.5.

[McKee, 1990b] observes that the competition multigraphs of the standard food web examples are not even chordal, let alone interval multigraphs (as they should be for a one-dimensional niche space). [McKee, 1995a] considers other deficiencies of food web models and uses competition multigraphs (and pseudographs) to predict possible omissions in observed food webs.

[McKee, 1994] introduces *intersection pseudographs*, formed by creating $|S_i|$ parallel loops at each vertex $S_i$ in $M \cong \Omega_\mu(\{S_1, \ldots, S_n\})$. When $|S_i| \geq 1$, there is a *multiple loop* at $S_i$ with *multiplicity* $|S_i|$; all such multiple loops are also included in $E(M)$.

**Example 6.3** Suppose $\mathcal{F} = \{S_1, \ldots, S_6\}$ where $S_1 = \{a\}$, $S_2 = \{e\}$, $S_3 = \{a, b, c\}$, $S_4 = \{a, b, c, d\}$, $S_5 = \{c, d, e, f\}$, and $S_6 = \{c, d, e, f\}$. Then the intersection pseudograph of $\mathcal{F}$ is shown on the left in Figure 6.4 with a set-labeled version on the right.

**Exercise 6.13** Show that every pseudograph is an intersection pseudograph.

We describe intersection pseudographs in further detail in order to illustrate concepts that are important in working with both intersection multigraphs and intersection pseudographs. A *maxclique $M$ of a pseudograph* is a

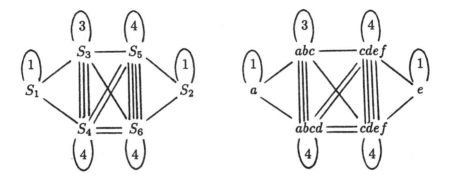

Figure 6.4: *An intersection pseudograph (with multiple loops).*

maxclique, together with one loop at each vertex, of the graph obtained by replacing each multiple edge and multiple loop in $E(M)$ by a simple edge or loop. Thus a maxclique contains exactly one edge and one loop from each bundle of parallel edges or loops having the same endpoints in $M$. If it is possible to remove simultaneously the edges and loops of all the max-cliques of $M$, then the resulting pseudograph is denoted $rM$. If, moreover, this process can be repeated, forming $r(rM)$, etc., until all edges and loops are gone, then $M$ is called a *reducible pseudograph* and the family (allowing repetitions) of all maxcliques of $M, rM, r(r(M)), \ldots$ are the *residual cliques* of $M$.

**Example 6.3 (continued)** The intersection pseudograph $M$ in Example 6.3 is reducible, with $rM$ and $r(r(M))$ shown in Figure 6.5. This $M$ has six residual cliques: $\{S_1, S_3, S_4\}$, $\{S_3, S_4, S_5, S_6\}$, and $\{S_2, S_5, S_6\}$, the max-cliques of $M$; $\{S_3, S_4\}$ and $\{S_4, S_5, S_6\}$, the maxcliques of $rM$; and $\{S_5, S_6\}$, the maxclique of $r(r(M))$.

**Exercise 6.14** Show that the intersection pseudograph of the family $\mathcal{F}$ in Example 6.2 is reducible, with eight residual cliques. Also show that the intersection pseudograph of the family $\mathcal{F}' = \{\{a\}, \{a, b\}, \{a, c\}, \{b, c\}\}$ is not reducible.

Define the *residual clique pseudograph* $K(M)$ of a reducible pseudograph $M$ to be the intersection pseudograph of all the residual cliques of $M$. The residual clique pseudograph of the pseudograph $M$ from Example 6.3 is shown in Figure 6.6. If $K(M)$ is also reducible and if $K(K(M)) \cong M$, then $K(M)$ is called the *pseudo dual* of $M$.

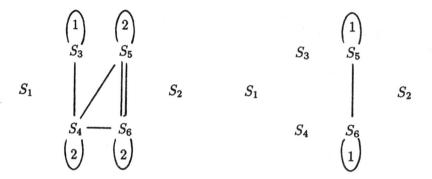

Figure 6.5: *Reduced pseudographs of Figure 6.4.*

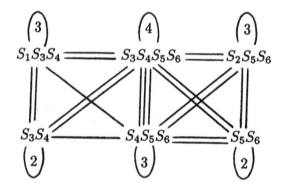

Figure 6.6: *The residual clique pseudograph of Figure 6.4.*

**Exercise 6.15** Show that the pseudograph in Example 6.3 has a pseudo dual.

**Exercise 6.16** Verify that the intersection pseudograph $M$ of the family $\mathcal{F} = \{S_1, \ldots, S_7\}$ with $S_1 = \{a, b, c, d\}$, $S_2 = \{a, b, c, e\}$, $S_3 = \{a, b, d, f\}$, $S_4 = \{a, c, d, g\}$, $S_5 = \{b, f\}$, $S_6 = \{d, g\}$, $S_7 = \{c, e\}$, is "self-dual" in the sense that $M \cong K(M)$.

[McKee, 1994] characterizes those pseudographs that have pseudo duals. Moreover, by defining *interval pseudographs* to be intersection pseudographs of subpaths of paths, every interval pseudograph can be shown to have a pseudo dual. [Fulkerson & Gross, 1965] shows algebraically, and [Duchet, 1984] states set-theoretically, that interval pseudographs have subpath representations that are unique up to isomorphism.

By defining *competition pseudographs* to be intersection pseudographs of out-neighborhoods of digraphs, every competition pseudograph of an *acyclic* digraph that also is an interval pseudograph can be shown to have a pseudo dual—namely, the digraph's "common-enemy pseudograph," paralleling section 4.2. [McKee, 1995a] relates this to the ecological application in section 4.3.

## 6.3 Tolerance Intersection Graphs

Suppose $\mathcal{F} = \{S_1, \ldots, S_n\}$ is a family of subsets of a finite set $S$, $\phi$ is a symmetric binary function taking pairs of positive reals to nonnegative reals, $\mu$ is a unary function taking subsets of $S$ to nonnegative reals, and each $S_i$ is assigned a positive real *tolerance $t_i$*. The *$\phi$-tolerance intersection graph $G$* of the family $\mathcal{F}$ with respect to $\phi$, $\mu$, and the $t_i$'s has vertex set $V(G) = \mathcal{F}$ with $S_i S_j \in E(G)$ if and only if $i \neq j$ and $\mu(S_i \cap S_j) \geq \phi(t_i, t_j)$. A graph $G$ is a *$\phi$-tolerance intersection graph* if there exist $\mathcal{F}$, $\phi$, $\mu$, and $t_i$'s such that $G$ is isomorphic to the $\phi$-tolerance intersection graph of $\mathcal{F}$ with respect to $\phi$, $\mu$, and the $t_i$'s.

This very general notion of $\phi$-tolerance was introduced in [Jacobson, McMorris, & Mulder, 1991] and [Jacobson, McMorris, & Scheinerman, 1991]. Frequently, $\mu$ measures the cardinality of a set or the length of an interval. Natural basic choices for the $t_i$'s include making them all a constant or setting each $t_i = \mu(S_i)$. Natural choices for $\phi$ include the minimum, maximum, product, sum, and absolute difference functions, resulting in the following types of tolerance graphs: *min-tolerance intersection graphs* from taking $\phi(x, y) = \min\{x, y\}$, *max-tolerance intersection graphs* from $\phi(x, y) = \max\{x, y\}$, *product-tolerance intersection graphs* from $\phi(x, y) = xy$, *sum-tolerance intersection graphs* from $\phi(x, y) = x + y$, and *abdiff-tolerance intersection graphs* from $\phi(x, y) = |x - y|$. The following exercise presents one extreme case of $\phi$-tolerance intersection graphs.

**Exercise 6.17** Show that if $\mu(S') = |S'|$ for each $S' \subseteq S$, each $t_i = \mu(S_i)$, and $\phi(x, y) = p$ (a positive integer) is a constant function, then the $\phi$-tolerance intersection graphs of a family $\mathcal{F}$ are precisely the $p$-intersection graphs of $\mathcal{F}$.

**Exercise 6.18** Show that every graph is a $\phi$-tolerance intersection graph for every $\phi$.

**Theorem 6.6 (Jacobson, McMorris, & Mulder)** *Every graph is a min-tolerance intersection graph of substars of a star with all the $t_i$'s equal to a constant.*

**Proof.** Let $G$ be a graph and $m = |E(G)|$. Label the pendant vertices of the star $K_{1,m}$ with the elements of $E(G)$. For each $v \in V(G)$, let $S_v$ denote the substar of $K_{1,m}$ that consists of the central vertex together with those pendant vertices labeled with the edges that are incident with $v$ in $G$. Thus $uv \in E(G)$ if and only if $S_u$ and $S_v$ have an edge of $K_{1,m}$ in common, namely the edge between the central vertex and the pendant vertex labeled $uv$. Thus $G$ is the min-tolerance intersection graph of the substars $S_v$ of $K_{1,m}$ with each $t_i = 2$ and $\mu(S_i) = |S_i|$.                    □

Note that Theorem 6.6 shows that every graph is the "edge intersection graph" of substars of a star, as shown in [Golumbic & Jamison, 1985a, 1985b]; also compare this with Corollary 2.17.

The definition of $\phi$-tolerance intersection graphs was motivated by the following special case from [Golumbic & Monma, 1982] and [Golumbic, Monma, & Trotter, 1984]. A *min-tolerance interval graph* is a min-tolerance intersection graph of a family of intervals of $\mathbb{R}$ in which $\mu$ measures the length of intervals. (Warning: in the literature, min-tolerance interval graphs are frequently referred to simply as "tolerance graphs.") In other words, the vertices of $G$ correspond to intervals $S_1, \ldots, S_n$ with two vertices $S_i$ and $S_j$ adjacent in $G$ if and only if $|S_i \cap S_j| \geq \min\{t_i, t_j\}$ for the corresponding intervals. [Golumbic, Monma, & Trotter, 1984] discusses possible applications that involve tolerating certain degrees of overlap of intervals.

**Exercise 6.19** Show that the cycle $C_4$, although not an interval graph, is a min-tolerance interval graph.

**Exercise 6.20** (see [Golumbic & Monma, 1982]) Show that if all the $t_i$'s equal a constant $c$, then the resulting min-tolerance interval graphs are interval graphs. Conversely, for every interval graph $G$, show that there is a constant $c$ such that $G$ is a min-tolerance interval graph with every $t_i = c$. (Using the results on "containment graphs" in section 7.6, if each $t_i = |S_i|$, then the resulting min-tolerance interval graphs can be shown to be "permutation graphs" as in section 7.6, and conversely.)

A wide variety of papers have been written on min-tolerance interval graphs, including [Monma, Reed, & Trotter, 1988], [Narasimhan & Manber, 1992], [Andreae, Hennig, & Parra, 1993], [Felsner, 1993], and [Holm & Bogart, to appear].

[Jacobson, McMorris, & Scheinerman, 1991] investigates *max-*, *sum-*, and *product-tolerance interval graphs*, where these are defined in the same sort of way as min-tolerance interval graphs.

**Theorem 6.7 (Jacobson, McMorris, & Scheinerman)** *Every tree is a max-tolerance interval graph.*

**Proof.** The proof is by induction and shows something even stronger: every tree is a max-tolerance interval graph such that, for each vertex, there is a representation in which the interval corresponding to that vertex is "left-most," meaning that its left-hand endpoint is less than or equal to all the other left-hand endpoints of intervals in the representation. To start the induction, note that such representations exist for trees having only one or two vertices.

Now let $T$ be a tree with $n$ vertices and assume that all trees with fewer than $n$ vertices are max-tolerance interval graphs where each vertex can be made to correspond to a left-most interval. Fix any $x \in V(T)$. Let $T_1, \ldots, T_k$ be the connected components of $T - x$ and let $x_i$ be the vertex of $T_i$ adjacent to $x$ in $T$. By the induction hypothesis, each $T_i$ has a max-tolerance interval representation with $x_i$ corresponding to a left-most interval. Note that, for each $i$, all the intervals in a representation of $T_i$ can be translated so that the left-hand endpoint of each left-most interval is 0. Assume that this has been done. Let $r_i$ denote the largest *right*-hand endpoint of this representation for $T_i$, and let $m$ be the length of the longest interval over all of these representations for $T_1, \ldots, T_k$. Choose $t' > m$ and set $m' = \max(t', t_{x_1}, \ldots, t_{x_k})$. Now extend the left-hand endpoint of each interval corresponding to an $x_i$ down to $-m'$. Each of these intervals is now $m'$ units longer and there are no new intersections among the $T_i$'s. Set $w_i = \sum_{j=1}^{i} r_j$.

We now define a max-tolerance interval representation of $T$ having $x$ correspond to a left-most interval. Let $S_x = [0, w_k + 2km']$ and $t_x = t'$, with the remaining tolerances as before and all the intervals in the $T_i$'s now translated $w_i + im$ units to the right. The new intervals corresponding to vertices in $T_i$ do not intersect those corresponding to vertices in $T_j$ for $i \neq j$ because they have been moved sufficiently far apart in the translations to the right. Thus we only need check those adjacencies caused by intervals intersecting $S_x$. Since $t' > m$ and each interval not corresponding to $x$ or an $x_i$ has length at most $m$, vertex $x$ can only possibly be adjacent to an $x_i$. Each $|S_{x_i}| \geq m'$, so $|S_x \cap S_{x_i}| \geq m' \geq \max(t_x, t_{x_i})$, and so $x$ is adjacent to each of $x_1, \ldots, x_k$. □

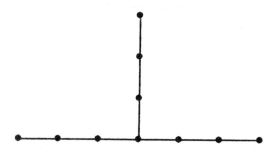

Figure 6.7: *A tree that is not a min-tolerance interval graph.*

**Exercise 6.21 (Jacobson, McMorris, & Scheinerman)** Show that every tree is a sum-tolerance interval graph and a product-tolerance interval graph. (Note that the same argument would work whenever $\phi$ satisfies $\lim_{x \to \infty} \phi(x, y) = \infty$.)

In contrast to these results, [Golumbic, Monma, & Trotter, 1984] shows that not every tree is a min-tolerance interval graph—indeed that a tree is a min-tolerance interval graph if and only if it contains no subtree isomorphic to the tree shown in Figure 6.7.

Motivated by Theorem 3.8 from [Roberts, 1969a], that the proper interval graphs are precisely the unit interval graphs, a natural question is whether every $\phi$-tolerance proper interval graph is a $\phi$-tolerance unit interval graph. As expected, a *$\phi$-tolerance proper interval graph* is a $\phi$-tolerance interval graph of a family of intervals where no interval is properly contained in another, and a *$\phi$-tolerance unit interval graph* is a $\phi$-tolerance interval graph of a family of unit-length intervals. Golumbic, Monma, and Trotter first posed this question for min-tolerance interval graphs and it was surprisingly answered in the negative in [Bogart, Fishburn, Isaak, & Langley, 1995]. The question remains open for max-tolerance. [Shull & Trenk, 1997] proves the equivalence of "unit" and "proper" for *bitolerance interval digraphs*, where the two endpoints are allowed to have different tolerances. [Jacobson & McMorris, 1991] answers the question in the affirmative for sum-tolerance interval graphs. This result will appear as Theorem 6.11, after we consider a couple of closely related classes of tolerance graphs.

[Bogart, Fishburn, Isaak, & Langley, 1995] defines a *50% tolerance graph* to be a min-tolerance interval graph represented by the intervals $S_i$ with tolerances $t_i = |S_i|/2$.

**Theorem 6.8 (Bogart, Fishburn, Isaak, & Langley)** *A graph is a 50% tolerance graph if and only if it is a min-tolerance unit interval graph.*

**Proof.** Suppose $G$ has a min-tolerance unit interval representation using the unit intervals $S_i$ and tolerances $t_i$. Note that each $t_i$ can be picked so that $t_i < |S_i| = 1$. Let $c_i$ be the center of $S_i$ and set $t'_i = 1 - t_i$. Create the new intervals $S'_i = [c_i - t'_i, c_i + t'_i]$ with tolerances $t'_i$. These intervals and tolerances give a representation for $G$ as a 50% tolerance graph.

Conversely, suppose $G$ has a 50% tolerance representation using intervals $S'_i$ and tolerances $t'_i = |S'_i|/2$. Scale the representation so that $t'_i < 1$ for all $i$ and let $c_i$ then be the center of the $i$th interval. Create the new intervals $S_i = [c_i - \frac{1}{2}, c_i + \frac{1}{2}]$ with tolerances $t_i = 1 - t'_i$. This new representation makes $G$ a min-tolerance unit interval graph. □

**Exercise 6.22** Verify the last step in each paragraph of the proof of Theorem 6.8.

[Jacobson, McMorris, & Mulder, 1991] and [Jacobson, Lehel, & Lesniak, 1993] study the $\phi$-*tolerance chain graphs* in which the family $S_1, \ldots, S_n$ is a chain of finite sets with respect to set inclusion, with $S_1 \subseteq \cdots \subseteq S_n$ and $\mu$ measuring cardinality.

**Exercise 6.23** Show that a graph is a $\phi$-tolerance chain graph if and only if it is the $\phi$-tolerance intersection graph of the sets $S_i = \{1, \ldots, k_i\}$ where each $k_i$ is an integer and $1 \leq k_1 \leq \cdots \leq k_n$.

**Exercise 6.24** Show that a graph is a $\phi$-tolerance chain graph if and only if it is a $\phi$-tolerance interval graph where the intervals $S_i = [0, r_i]$ where each $r_i$ is real and $0 < r_1 \leq r_2 \leq \cdots \leq r_n$.

The next exercise demonstrates what happens when the two natural restrictions are placed on the tolerances and the measure $\mu$.

**Exercise 6.25** Show the following:

(1) A graph is a $\phi$-tolerance chain graph with constant tolerances if and only if it consists of a complete graph and isolated vertices.

(2) A graph is a min-tolerance chain graph with tolerances equal to set-sizes if and only if it is a complete graph.

(3) A graph is a max-tolerance chain graph with tolerances equal to set-sizes if and only if it is a disjoint union of complete graphs.

(4) A graph is a sum-tolerance chain graph with tolerances equal to set-sizes if and only if it is an edgeless graph.

We now introduce some convenient notation. If $G$ is the $\phi$-tolerance interval graph of the intervals $[l_i, r_i]$ with corresponding tolerances $t_i$ for $i = 1, \ldots, n$, we simply say that $G$ has the *representation* $[l_1, r_1]; t_1, \ldots,$ $[l_n, r_n]; t_n$ (or $[l_i, r_i]; t_i$). Exercise 6.24 shows that a graph is a $\phi$-tolerance chain graph if and only if it has a representation $[0, r_1]; t_1, \ldots, [0, r_n]; t_n$, where $0 < r_1 \leq \cdots \leq r_n$. The following lemma is easy to prove.

**Lemma 6.9** *Suppose $G$ is a $\phi$-tolerance interval graph.*

*(1) If $G$ has the representation $[l_i, r_i]; t_i$, then $G$ has the representation $[l_i + k, r_i + k]; t_i$ for every $k$.*

*(2) Suppose $\phi$ satisfies the condition that $\phi(kx, ky) = k\phi(x, y)$ for all positive $k$. Then $G$ has the representation $[l_i, r_i]; t_i$ if and only if $G$ has the representation $[kl_j, kr_j]; kt_i$ for every positive $k$.* □

**Theorem 6.10 (Jacobson & McMorris)** *Every $\phi$-tolerance chain graph is a $\phi$-tolerance proper interval graph.*

**Proof.** Let $G = (V, E)$ be a $\phi$-tolerance chain graph with representation $[0, r_i]; t_i$ with $r_i \leq r_j$ whenever $i \leq j$. If $G$ is complete we are done. If not, then let $\epsilon = \min\{\phi(t_i, t_j) - \min(r_i, r_j) : v_i v_j \notin E\}$. Since $G$ is not complete, $\epsilon > 0$. Now set $S_j = \left[\frac{-(n-j)\epsilon}{2n}, r_j + \frac{j\epsilon}{2n}\right]$. Clearly $|S_j| = r_j + \frac{\epsilon}{2}$. It follows that $|S_i \cap S_j| \geq \phi(t_i, t_j)$ if and only if $\min[r_i, r_j] \geq \phi(t_i, t_j)$. Therefore, $S_1; t_1, \ldots, S_n; t_n$ is a proper interval representation of $G$. □

**Theorem 6.11 (Jacobson & McMorris)** *A graph is a sum-tolerance proper interval graph if and only if it is a sum-tolerance unit interval graph.*

**Proof.** It is easy to see that every sum-tolerance unit interval graph is a sum-tolerance proper interval graph. Let $G$ be a sum-tolerance proper interval graph having representation $[l_1, r_1]; t_1, \ldots, [l_n, r_n]; t_n$. By Lemma 6.9 we are done if we show that $G$ has a representation using intervals of equal length. Using induction, assume that the first $k$ intervals have equal length. If $k = n$ we are done, so assume that $k < n$.

We now show how to construct a representation for $G$ where the first $k+1$ intervals all have the same length. Let $S_i = [l_i, r_i]$. Assume $|S_k| > |S_{k+1}|$ and set $\delta = |S_k| - |S_{k+1}|$. Form the intervals $S_j'$ as follows: For $j \leq k$ let $S_j' = S_j$, and for $j > k$ let $S_j' = [l_j - \frac{\delta}{2}, r_j + \frac{\delta}{2}]$. Let $t_j' = t_j$ for all $j \leq k$ and $t_j' = t_j + \frac{\delta}{2}$ for $j > k$. It is easy to show that $S_j'; t_j'$ is a representation for $G$ for each $j = 1, \ldots, n$. Similarly if $|S_k| < |S_{k+1}|$, let $\delta = |S_{k+1}| - |S_k|$ and set $S_j' = S_j$ for $j > k$. For $j \leq k$, let $S_j' = [l_j - \frac{\delta}{2}, r_j + \frac{\delta}{2}]$ and define the new tolerances by $t'_j = t_j$ for $j \leq k$ and $t'_j = t_j + \frac{\delta}{2}$ for $j > k$. $\quad\square$

The next three theorems demonstrate the robustness of $\phi$-tolerance chain graphs.

**Theorem 6.12 (Jacobson & McMorris)** *A graph is a sum-tolerance proper interval graph if and only if it is a sum-tolerance chain graph.*

**Proof.** From Theorem 6.10, every sum-tolerance chain graph is a sum-tolerance proper interval graph. To prove the converse, let $G$ be a sum-tolerance proper interval graph with representation $[l_1, r_1]; t_1, \ldots, [l_n, r_n]; t_n$. Since the intervals are proper we may assume that $l_1 < \cdots < l_n$ and $r_1 < \cdots < r_n$. By Lemma 6.9, we can take $0 \leq l_1$. We now show that $G$ is a sum-tolerance chain graph with representation $[0, r_1 + l_1]; t_1 + l_1, \ldots, [0, r_n + l_n]; t_n + l_n$: For $i < j$, $|[0, r_i + l_i] \cap [0, r_j + l_j]| \geq (t_i + l_i) + (t_j + l_j)$ if and only if $r_i + l_i \geq t_i + l_i + t_j + l_j$ if and only if $r_i - l_j \geq t_i + t_j$ if and only if $|[l_i, r_i] \cap [l_j, r_j]| \geq t_i + t_j$. $\quad\square$

**Theorem 6.13 (Jacobson, McMorris, & Mulder)** *A graph is a max-tolerance chain graph if and only if it is an interval graph.*

**Proof.** Suppose $G$ is a max-tolerance chain graph with, by Exercise 6.23, each $S_i = \{1, \ldots, k_i\}$ where $1 \leq k_1 \leq \cdots \leq k_n$. Consequently, vertices corresponding to $S_i$ and $S_j$ are adjacent in $G$ if and only if $\min\{k_i, k_j\} \geq \max\{t_i, t_j\}$. We may assume that each $t_i \leq k_i$, since if $t_i > k_i$, then $S_i$ corresponds to an isolated vertex of $G$ and can be disregarded. So vertices corresponding to $S_i$ and $S_j$ are adjacent if and only if $[t_i, k_i] \cap [t_j, k_j] \neq \emptyset$. Thus, the max-tolerance chain graph of $N$ is the interval graph of the set $\{[t_i, k_i] : i = 1, \ldots, n\}$. Conversely, it is easy to show that every interval graph has an interval representation of this form. $\quad\square$

**Theorem 6.14 (Jacobson, McMorris, & Mulder)** *A graph is a min-tolerance chain graph if and only if it is a threshold graph.*

**Proof.** Suppose $G$ is a min-tolerance chain graph with a representation $[0, r_1]; t_1, \ldots, [0, r_n]; t_n$ where $r_1 \leq \cdots \leq r_n$ with vertex $v_i$ corresponding to $[0, r_i]$. By Theorem 5.1, it suffices to show that every such $G$ either has a vertex adjacent to all other vertices or has an isolated vertex. If $t_1 \leq r_1$, then $v_1$ is adjacent to all other vertices of $G$. If $t_1 > r_1$ and $v_1$ is not isolated, let $v_j$ be adjacent to $v_1$. This implies that $t_j \leq r_1$, and thus $v_j$ is adjacent to all other vertices of $G$.

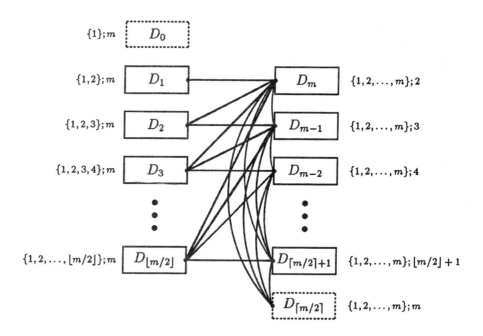

Figure 6.8: *A threshold graph with a min-tolerance representation for the proof of Theorem 6.14.*

For the converse, Figure 6.8 shows a min-tolerance representation for a typical threshold graph $G$ (as in Figure 5.3): $D_0, \ldots, D_m$ is the degree partition of $G$ with $D_0$ possibly empty and $D_{\lceil m/2 \rceil}$ present only if $m$ is odd. A line between cells $D_i$ and $D_j$ means that every vertex in $D_i$ is adjacent to every vertex in $D_j$. The $D_i$'s in the left column represent independent sets, with the open neighborhoods of their vertices ordered by inclusion downward, and the $D_i$'s in the right column represent complete subgraphs, with the closed neighborhoods of their vertices ordered by inclusion upward. The representing set and tolerance are next to each $D_i$. □

There are other interesting ways to extend the notion of threshold graphs. [Jacobson, Lehel, & Lesniak, 1993] calls a graph $G$ a *$\phi$-threshold graph* if there exists a positive number $c$ assigned to $G$ and a positive weight $w_v$ assigned to each vertex $v$ such that

$$uv \in E(G) \text{ if and only if } c \geq \phi(w_u, w_v).$$

Since the complement of a threshold graph is a threshold graph by Corollary 5.2, the ordinary threshold graphs are precisely the sum-threshold graphs.  [Jacobson, Lehel, & Lesniak, 1993] goes on to characterize $\phi$-threshold graphs where $\phi = a(x+y) + b|x-y|$.

[Monma, Reed, & Trotter, 1988] defines a graph $G$ to be a *threshold tolerance graph* if it is possible to assign a positive weight $w_v$ to each vertex $v$ of $G$ and a positive tolerance $t_v$ to each $v$ such that

$$uv \in E(G) \text{ if and only if } w_u + w_v > \min(t_u, t_v).$$

Thus if all the tolerances are equal in a threshold tolerance graph, it is an ordinary threshold graph.  Complements of threshold tolerance graphs, called *coTT graphs*, are characterized in [Monma, Reed, & Trotter, 1988].  Using the same parameters as above, adjacency in a coTT graph is defined by

$$w_u + w_v \leq \min(t_u, t_v).$$

By changing the $t_v$'s to $r_v$'s and then the $w_v$'s to $t_v$'s, it is clear from Exercise 6.24 that coTT graphs are precisely sum-tolerance chain graphs.  The following summarizes these results on sum-tolerance chain graphs.

**Theorem 6.15 (Jacobson, McMorris, & Mulder)** *Let $G$ be a graph. Then the following statements are equivalent:*

*(1) $G$ is a coTT graph;*

*(2) $G$ is a sum-tolerance chain graph;*

*(3) $G$ is a sum-tolerance unit interval graph;*

*(4) $G$ is a sum-tolerance proper interval graph.*                    $\square$

[Brigham, McMorris, & Vitray, 1995] studies *$\phi$-tolerance competition graphs*, defined the same way that competition graphs and $p$-competition graphs were in sections 4.2 and 6.1.  Specifically, let $\phi$ be a symmetric binary function that takes pairs of nonnegative integers to nonnegative integers.  The graph $G$ is a $\phi$-tolerance competition graph if $G$ is isomorphic to the $\phi$-tolerance intersection graph of the out-neighborhoods of the vertices of some digraph.  Thus

$$v_i v_j \in E(G) \text{ if and only if } |N^+(v_i) \cap N^+(v_j)| \geq \phi(t_i, t_j)$$

for some tolerances $t_1, \ldots, t_n$.  As expected, there is a clique cover analogue of Theorem 6.3.  Let $\phi$ be as above and $T = (t_1, \ldots, t_n)$ be an $n$-tuple of (not necessarily distinct) nonnegative integers.  A *$\phi$-$T$-edge clique cover* of the graph $G$ is a family $\{V_1, \ldots, V_k\}$ of subsets of $V(G)$ such that $v_i v_j \in E(G)$ if and only if $v_i$ and $v_j$ are elements of at least $\phi(t_i, t_j)$ common sets $V_l$.

**Exercise 6.26 (Brigham, McMorris, & Vitray)** *Show that a graph G is a $\phi$-tolerance competition graph if and only if there is a $\phi$-T-edge clique cover of G with $|V|$ elements.*

While there are many results concerning various $\phi$-tolerance competition graphs, major questions remain open. It is not even known whether there are graphs (possibly even tripartite graphs) that are not min-tolerance competition graphs. Related papers include [Anderson, Langley, Lundgren, McKenna, & Merz, 1994] and [Brigham, McMorris, & Vitray, 1995]. [Brigham, Carrington, & Vitray, to appear] introduces abdiff-tolerance competition graphs and characterizes those complete bipartite graphs that are abdiff-tolerance competition graphs.

# Chapter 7

# Guide to Related Topics

This chapter consists of sections involving clusters of concepts related to intersection graphs. Arranged alphabetically, they can be read in any order and the index should help in navigating between related topics.

Each section contains selected definitions and states results without proof, concentrating on the flavor of the topic and pointers to both the original papers and recent work, especially to surveys with extensive bibliographies. [Brandstädt, 1993] contains more information on many of the families of graphs considered (and [Brandstädt, Le, & Spinrad, to appear] will surely contain much more).

We are well aware that we have not covered many topics in which work is being done, and that we have not given complete coverage of any of these topics—these are all areas of active research. Also, while many of these families are widely studied in terms of the computational complexity of problems like domination and coloring, our attention is concentrated on structural aspects.

## 7.1 Assorted Geometric Intersection Graphs

It is clearly possible, and sometimes useful, to consider intersection graphs of all sorts of geometric objects; occasionally, nice results have surfaced. While we are admittedly spotty in our treatment of the array of possibilities, we emphasize what seem to be natural examples, yet try to mention a few of the unexpected (for instance, [Maire, 1993] studies the intersection graphs of maximal rectangles in polyominoes).

Motivated by the success of interval graphs, and since intervals are the convex subsets of $\mathbb{R}^1$, [Wegner, 1967] shows that $K_5$ with each edge bisected

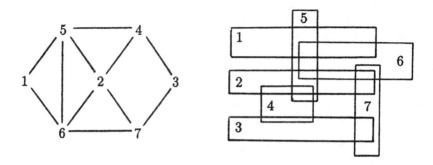

Figure 7.1: *A boxicity two graph with a 2-box representation.*

is an example of a graph that is not an intersection graph of convex subsets of $\mathbb{R}^2$. The problem of characterizing those graphs that are intersection graphs of convex subsets of $\mathbb{R}^2$ is still open, but the following two results from [Duchet, 1978, 1984] and [Wegner, 1967] are known.

**Theorem 7.1 (Duchet)** *Every chordal graph is the intersection graph of convex subgraphs of the plane.*

**Theorem 7.2 (Wegner)** *Every graph is the intersection graph of convex subsets of $\mathbb{R}^3$.*

A *d-dimensional box* is the cartesian product of intervals $[a_i, b_i]$ for $1 \leq i \leq d$. A graph is a *d-box graph* if it is the intersection graph of $d$-dimensional boxes in $\mathbb{R}^d$. Hence interval graphs are precisely the 1-box graphs. Just as interval graphs can also be viewed as intersection graphs of subpaths of paths, $d$-box graphs can be viewed as intersection graphs of subgrids of *d-dimensional grids*: $d$-dimensional integer lattice points, with two points adjacent if and only if they differ by 1 at one coordinate position and not at all at the other positions.

[Roberts, 1969b] observes that every graph of order $n$ is an $n$-box graph. Thus the *boxicity* of a graph $G$ can be defined to be the minimum $d$ for which $G$ is a $d$-box graph. Figure 7.1 shows an example of a graph of boxicity two.

**Theorem 7.3 (Roberts)** *Every complete multipartite graph $K_{n_1,...,n_p}$ has boxicity equal to $|\{i : n_i > 1\}|$.*

Hence $C_4 = K_{2,2}$ has boxicity two and the octahedron $K_{2,2,2}$ has boxicity three, showing that the following results from [Scheinerman, 1984] and [Thomassen, 1986] are best possible. (Recall that a graph is *outerplanar* if it can be embedded in the plane with all of its vertices on the unbounded face.)

**Theorem 7.4 (Scheinerman)** *Every outerplanar graph has boxicity at most two.*

**Theorem 7.5 (Thomassen)** *Every planar graph has boxicity at most three.*

[Roberts, 1989] contains a short survey of known results about $d$-box graphs and boxicity, including NP-completeness and other references, but surprisingly little is known in general. [Quest & Wegner, 1990] gives a matrix-based characterization of graphs that have boxicity at most two, and [Rim & Nakajima, 1995] discusses computational problems on 2-box graphs. [Trotter & West, 1987] presents a related notion of representability, replacing boxes in $d$-space with intervals in $d$-dimensional partially ordered sets.

Generalizing $d$-box graphs—$d$-dimensional boxes in $d$-dimensional grids—[Hartman, Newman, & Ziv, 1991] and [Bellantoni, Hartman, Przytycka, & Whitesides, 1993] define the *grid dimension* of a graph to be the smallest $k$ such that the graph is the intersection graph of $d$-dimensional boxes in a $(d + 1)$-dimensional grid.

**Theorem 7.6 (Bellantoni, Hartman, Przytycka, & Whitesides)**
*Every graph has grid dimension equal to or one less than its boxity.*
*A bipartite graph has boxicity less than or equal to two if and only if it has grid dimension less than or equal to one.*

[Maehara, 1984b] and [Erdős, Godsil, Krantz, & Parsons, 1988] study intersections graphs of balls in $\mathbb{R}^n$, and [Sachs, 1994] studies their tangency graphs. [Clark, Colbourn, & Johnson, 1990] and [Marathe, Breu, Hunt, Ravi, & Rosenkrantz, 1995] consider computational problems on the special case of *unit disk graphs*, an intersection class with obvious applications to cellular telephone networks.

A *circular-arc graph* $G$ is isomorphic to the intersection graph of a family of closed arcs of a circle or, equivalently, of a family of connected subgraphs of a cycle. Interval graphs clearly are circular-arc graphs, but circular-arc graphs are, in general, very different from interval graphs since they need not

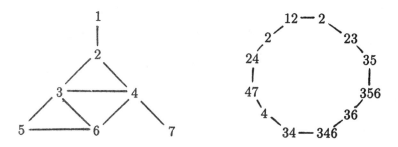

Figure 7.2: *A circular-arc graph with corresponding subpaths in a cycle.*

even be chordal: every cycle $C_n$ is a circular-arc graph. Figure 7.2 shows a
circular-arc graph with a subpaths-of-a-cycle intersection representation that
can easily be smoothed into an arcs-of-a-circle representation. Section 8.6
of [Golumbic, 1980], [Tucker, 1978], and [Flotow, 1996] discuss applications
of circular-arc graphs, and [Gavril, 1974b] is a key paper. Recognition algorithms for circular-arc graphs are discussed in [Eschen & Spinrad, 1993]
(using chordal bipartite graphs), [Hsu, 1995], and [Hell & Huang, 1997].
[Kloks, Kratsch, & Wong, 1996] gives a cubic algorithm for the minimum
fill-in problem.

Circular-arc graphs do not have the graph-theoretic characterizations
that might be expected from their superficial resemblance to interval graphs,
largely due to arcs (those in Figure 7.2 corresponding to vertices 2, 3, and
4, for instance) of a circle not having to satisfy the Helly condition.

Recall the maxclique-vertex matrix $M(G)$ of a graph $G$ defined in section 3.3, and that $G$ is an interval graph if and only if $M(G)$ has the consecutive ones property for columns. Define $M(G)$ to have the *circular ones
property for columns* if all the 1 entries in each column are consecutive when
the matrix is thought of as wrapped around a horizontal cylinder. If $M(G)$
has the circular ones property for columns, then $G$ is a circular-arc graph,
but the example in Figure 7.2 shows that the converse fails. [Gavril, 1974b]
defines a graph to be a *Helly circular-arc graph* if it is isomorphic to the
intersection graph of a family of arcs of a circle that satisfies the Helly condition, and proves that $M(G)$ has the circular ones property for columns
if and only if $G$ is a Helly circular-arc graph.  Chapter 6 of [Golumbic,
1980] and [Cozzens & Mahadev, 1989] contain more on the consecutive ones
property and its generalizations.

Define the *augmented adjacency matrix* $A^+(G)$ of a graph $G$ to be its
adjacency matrix with each diagonal entry set equal to 1. [Roberts, 1968]

proves that $G$ is a proper interval graph if and only if $A^+(G)$ has the consecutive ones property for columns. [Tucker, 1971] proves that, if $A^+(G)$ has the circular ones property for columns, then $G$ is a circular-arc graph, but the example in Figure 7.2 again shows that the converse fails. Tucker does characterize circular-arc graphs by $A^+(G)$ satisfying a "quasi circular ones property."

[Tucker, 1971, 1974] give characterizations of *proper circular-arc graphs*—graphs with a circular-arc representation in which none of the arcs properly contains another—and *unit circular-arc graphs*—graphs with a circular-arc representation of equal-length arcs. Unlike what happened in section 3.3, unit circular-arc graphs form a proper subclass of the proper circular-arc graphs. The nonintersection of chords that subtend nested arcs of a circle leads to the following connection with circle graphs (section 7.4).

**Theorem 7.7** *Every proper circular-arc graph is a circle graph.*

The following characterization of proper circular-arc graphs is in [Skrien, 1982]. (The same statement can be used to characterize proper interval graphs by also requiring the orientation to be acyclic.)

**Theorem 7.8 (Skrien)** *A connected graph is a proper circular-arc graph if and only if it has an orientation that contains no induced subdigraph isomorphic to* $\bullet\!\longrightarrow\!\bullet\!\longleftarrow\!\bullet$ *or* $\bullet\!\longleftarrow\!\bullet\!\longrightarrow\!\bullet$.

[Hell & Huang, 1995] and [Deng, Hell, & Huang, 1996] contain proper circular-arc graph algorithms. [Stueckle, Piazza, & Ringeisen, 1995] contains an application of proper circular-arc graphs to questions involving how graphs can be drawn. [Bang-Jensen & Hell, 1994] discusses chordal proper circular-arc graphs.

Stepping up from $\mathbb{R}^1$, [Ehrlich, Even, & Tarjan, 1976] and [Kratochvíl & Matoušek, 1994] study the intersection graphs of line segments in $\mathbb{R}^2$, showing that recognizing such graphs is NP-hard even when all the segments lie in a fixed number (greater than one) of directions. Section 7.4 discusses several other special cases that are more nicely behaved.

[Ehrlich, Even, & Tarjan, 1976] and [Kratochvíl, 1991a, 1991b] study *string graphs*, the intersection graphs of arbitrary curves in $\mathbb{R}^2$. If the curves are all "graphs" of continuous functions on the closed unit interval, then [Golumbic, Rotem, & Urrutia, 1983] shows that the intersection graphs are precisely the complements of the comparability graphs (section 7.6). On the other hand, every graph is the intersection graph of arbitrary curves in $\mathbb{R}^3$.

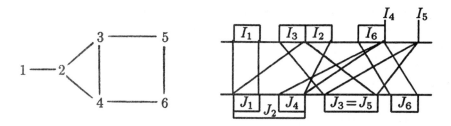

Figure 7.3: *A trapezoid graph with an intersection representation.*

[Corneil & Kamula, 1987] and [Degan, Golumbic, & Pinter, 1988] gener-
alize interval graphs in another way. Suppose $L$ and $M$ are two parallel lines
of which, respectively, $\{I_1, \ldots, I_n\}$ and $\{J_1, \ldots, J_n\}$ are families of intervals.
Then, each $i \in \{1, \ldots, n\}$ determines a trapezoid having parallel sides $I_i$
and $J_i$ (allowing degenerate trapezoids with either $I_i$ or $J_i$ a single point).
A graph is a *trapezoid graph* if it is isomorphic to the intersection graph of
such a family of trapezoids.

Every interval graph is a trapezoid graph, taking each pair $I_i$ and $J_i$ so as
to make the trapezoid a rectangle. Every permutation graph (section 7.4)
is a trapezoid graph, taking each $I_i = \{i\}$ and $J_i = \{\pi(i)\}$ where $\pi$ is a
permutation of $\{1, \ldots, n\}$. The graph shown in Figure 7.3, from [Corneil &
Kamula, 1987], shows a trapezoid graph that is neither an interval graph
nor a permutation graph. [Cheah & Corneil, 1996] contains more on the
structure of trapezoid graphs and their relation to permutation graphs.

[Flotow, 1995] introduces higher-dimensional analogues of trapezoid
graphs. [Felsner, Müller, & Wernisch, 1997] contains many relevant ideas,
including a more general notion of a "circle trapezoid graph" that subsumes
both circle graphs (section 7.4) and circular-arc graphs. The following two
theorems from [Corneil & Kamula, 1987] and [Felsner, 1993] link trapezoid
graphs with cocomparability graphs (section 7.6).

**Theorem 7.9 (Corneil)**   *Every trapezoid graph is a cocomparability
graph.*

**Theorem 7.10 (Kamula and Felsner)** *Every cocomparability graph
that is also a min-tolerance intersection graph is a trapezoid graph.*

[Trotter & Harary, 1978] and [Griggs & West, 1979] independently introduce a natural generalization of interval graphs by allowing each vertex
of a graph to be represented by a *union* of intervals. The *interval number*

of a graph $G$ is the smallest number $t$ such that $G$ is an intersection graph
with each vertex corresponding to a union of at most $t$ intervals. The interval graphs are, of course, precisely the graphs having interval number
one. Every cycle $C_n : v_1, v_2, \ldots, v_n, v_1$ with $n \geq 4$ clearly has interval number two, using two intervals to correspond to $v_1$ and one each for the rest.
[West & Shmoys, 1984] shows that recognizing graphs having a fixed interval
number is NP-complete. [Scheinerman & West, 1983] contains an extensive
discussion, including that $K_{2,9}$ has interval number three and the following
theorem.

**Theorem 7.11 (Scheinerman & West)** *Every planar graph has interval number at most three.*

The interval number of $G$ is bounded above by the maximum degree
of $G$, by $\lceil (|V(G)| + 1)/4 \rceil$ from [Griggs, 1979], and by $1 + \lceil \sqrt{|E(G)|/2} \rceil$
from [Spinrad, Vijayan, & West, 1987]. While chordal graphs can have
arbitrarily large interval numbers, [Scheinerman, 1988a] discusses specific
upper bounds. [Kratzke & West, 1993, 1996] discuss other parameters that
resemble interval numbers. [Joseph, Meidanis, & Tiwari, 1992] contains a
molecular biology application of graphs having interval number at most two.
[Raychaudhuri, 1992b] uses multiple interval assignments in a traffic phasing
context.

Partially motivated by [Kumar & Deo, 1994], [Gyárfás & West, 1995]
discusses *multitrack interval graphs* in which vertices correspond to intervals from separate copies of the real line ("parallel tracks"). [Scheinerman,
1985b] contains a very general treatment of multiple-set representations.

As we change our focus from interval graphs to chordal graphs, observe
that a graph $G$ having boxicity at most $d$ can be equivalently phrased as
saying that it is the "intersection" of interval graphs $G_1, \ldots, G_d$, where this
means that $V(G) = V(G_1) = \cdots = V(G_d)$ and $E(G) = E(G_1) \cap \cdots \cap E(G_d)$.
(The equivalence can be seen by viewing the projections of two-dimensional
boxes on the $x$- and $y$-axes as intervals.) Motivated by this, a graph is said to
have *chordality* at most $d$ if it is similarly an intersection of $d$ chordal graphs.
For instance, Figure 7.4 shows that $C_5$ is the intersection of two chordal
graphs and so has chordality two. The octahedron $K_{2,2,2}$ has chordality
three. Every bipartite graph has chordality at most two since it is the
intersection of two split graphs. The analogues of Theorems 7.4 and 7.5
hold since chordality is always less than or equal to boxicity. [Cozzens &
Roberts, 1989] and [McKee & Scheinerman, 1993] contain many bounds and
other results on chordality, of which we mention only one.

Figure 7.4: *A cycle as the intersection of two chordal graphs.*

A *k-tree* can be defined as a chordal graph that has a perfect elimination ordering $\langle v_1, \ldots, v_n \rangle$ such that each $v_i$ has degree $\min\{k, n-i\}$ in the subgraph induced by $v_i, \ldots, v_n$ (so the 2-trees are exactly the trees). The *treewidth* of $G$ is the least $k$ such that $G$ is a *partial k-tree*, meaning that $G$ is a subgraph of a $k$-tree. [Kloks, 1994] contains an extensive discussion of treewidth and other topics that have grown out of the pioneering work of Robertson and Seymour—see, for instance, [Robertson & Seymour, 1985]— including calculating and approximating the treewidth of many of the same families of graphs that we study.

**Theorem 7.12 (McKee & Scheinerman)** *Every graph has chordality less than or equal to its treewidth.*

Thus the *series-parallel graphs*—the partial 2-trees—have chordality at most two, in contrast to an example in [McKee & Scheinerman, 1993] of a series-parallel graph that has boxicity three.

[Kratochvíl & Tuza, 1994] and [Hliněný & Kuběna, 1995] discuss more general "intersection dimensions" for classes of graphs. Mimicking multi-track interval graphs, [Chang, Jacobson, Monma, & West, 1993] gives results involving unions of subtrees (or substars) of trees.

Motivated by chordal graphs being the intersection graphs of subtrees of trees, [Renz, 1970] characterizes the intersection graphs of sub*paths* of a tree, and [Gavril, 1978] gives a recognition algorithm. [Golumbic & Jamison, 1985a, 1985b] and [Sysło, 1985] investigate what happens when the paths are considered as sets of edges (rather than sets of vertices). [Monma & Wei, 1986] presents an extensive study of *path graphs*, including many sorts of intersection graphs involving various sorts of families of subpaths of a tree. Viewing each subpath as a set of vertices, there are three possible intersection classes: *UV*: intersection graphs of undirected paths of an undirected tree, *DV*: intersection graphs of directed paths of a directed tree, and *RDV*: intersection graphs of directed paths of a rooted directed tree; three other possible intersection classes—*UE*, *DE*, and *RDE*, respectively— result by viewing each subpath as a set of edges. In particular, Monma

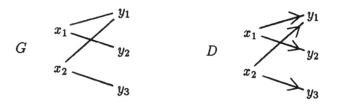

Figure 7.5: *A bipartite intersection graph $G$ from $S_{x_1} = \{a, b\}$, $S_{x_2} = \{c\}$, $T_{y_1} = \{b, c\}$, $T_{y_2} = \{a\}$, $T_{y_3} = \{c\}$, and its associated intersection digraph $D$.*

and Wei present clique tree characterizations of these classes, in the style of Theorem 2.1.

**Theorem 7.13 (Monma & Wei)** *The six intersection classes are distinct and related as follows*: RDV $\Rightarrow$ DV $\Rightarrow$ UV $\Rightarrow$ chordal, *and* RDE $\Rightarrow$ DE $\Rightarrow$ UE. *($C_5$ is an example of a graph in* UE *that is not chordal.)*

**Theorem 7.14 (Monma & Wei and Golumbic & Jamison)** *Membership in* UV, DV, RDV, DE, *or* RDE *can be recognized by a unified polynomial algorithm, but recognizing members of* UE *(or even recognizing whether a member of* UV *is in* UE*) is NP-complete.*

[Panda & Mohanty, 1995] discusses some of these classes further. [Gavril, 1994, 1996], [Gavril & Urrutia, 1994], and [Prisner, 1994] go another way from chordal graphs by looking at intersection graphs of various sorts of subtrees of classes of graphs that are more general than trees.

## 7.2   Bipartite Intersection Graphs, Intersection Digraphs, and Catch (Di)Graphs

[Harary, Kabell, & McMorris, 1982] defines a bipartite graph $G$ with $V(G) = X \cup Y$ and $X \cap Y = \emptyset$ to be a *bipartite intersection graph*, sometimes called an *intersection bigraph*, if each $x \in X$ can be assigned a set $S_x$ and each $y \in Y$ a set $T_y$ such that $xy \in E(G)$ if and only if $S_x \cap T_y \neq \emptyset$. Figure 7.5 gives an example.

[Sen, Das, Roy, & West, 1989] defines a directed graph $D$, with loops allowed, to be an *intersection digraph* if each $v \in V(D)$ can be assigned two sets $S_v$ and $T_v$ such that $uw \in A(D)$ if and only if $S_u \cap T_w \neq \emptyset$. (This is

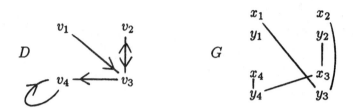

Figure 7.6: *An intersection digraph $D$ from $S_{v_1} = \{a\}$, $T_{v_1} = \{b, c\}$, $S_{v_2} = \{a\}$, $T_{v_2} = \{c, d\}$, $S_{v_3} = \{d\}$, $T_{v_3} = \{a\}$, $S_{v_4} = \{e\}$, $T_{v_4} = \{c, d, e\}$, and its associated bipartite intersection graph $G$.*

a specialization of the earlier notion of "connection digraph," introduced in [Beineke & Zamfirescu, 1982].) Figure 7.6 gives an example.

Bipartite intersection graphs and intersection digraphs are intimately interrelated, a feature that is frequently useful in their study. For instance, each bipartite intersection graph $G$ as above leads to an intersection digraph $D$ on $V(D) = X \cup Y$ by setting $T_x = \emptyset$ for each $x \in X$ and $S_y = \emptyset$ for each $y \in Y$; this means that $D$ results from $G$ by directing each edge $xy \in E(G)$ from $x \in X$ toward $y \in Y$. See Figure 7.5.

Conversely, each intersection digraph $D$ as above leads to a bipartite intersection graph $G$ on $V(G) = \{x_v, y_v : v \in V(D)\}$ by setting, for each $v \in V(D)$, $S_{x_v} = S_v$, $T_{x_v} = \emptyset$, $S_{y_v} = \emptyset$, and $T_{y_v} = T_v$; this means that $D$ results from $G$ by directing each edge $x_u y_w \in E(G)$ from $x_u$ toward $y_w$ and then identifying each $x_v, y_v$ pair. See Figure 7.6.

[Harary, Kabell, & McMorris, 1982] focuses on *bipartite interval graphs*, where each $S_x$ and $T_y$ is an interval of a line; [Müller, 1997] updates work toward a characterization. [Sen, Das, Roy, & West, 1989]—see also [West, 1998]—similarly focuses on *interval digraphs*, where each $S_v$ and $T_v$ is an interval of a line. Sen, Das, and West give an elegant adjacency matrix characterization of interval digraphs. The analogy between interval graphs $G$ and interval digraphs $D$ involves replacing the fundamental role of a complete subgraph $Q$ of $G$ with a subdigraph of $D$ formed from $X, Y \subseteq V(D)$ with $uw \in A(D)$ if and only if $u \in X$ and $w \in Y$, and with a loop at any vertex in $X \cap Y$. [Müller, 1997] gives a polynomial algorithm for recognizing interval digraphs, and so for recognizing bipartite interval graphs. [Langley, Lundgren, & Merz, 1995] studies the competition graphs of interval digraphs.

[Sen & Sanyal, 1994] introduces notions of *unit interval, proper interval,* and *indifference digraphs* as the natural modifications of the undirected notions from sections 3.3 and 3.4.2; see also [West, 1998].

**Theorem 7.15 (Sen & Sanyal)** *The properties of being a unit interval digraph, a proper interval digraph, or an indifference digraph are all equivalent.*

[Steiner, 1996] gives a linear recognition algorithm for the digraphs in Theorem 7.15; [Lin & West, 1995] characterizes them with forbidden submatrices. [Sanyal & Sen, 1996] studies other classes of interval digraphs.

[Sen, Das, & West, 1989, 1992] introduce *circular-arc digraphs*, in analogy with circular-arc graphs (section 7.1), and [Sen, Sanyal, & West, 1995] studies various other sorts of intersection digraphs, including directed containment graphs in analogy with section 7.6.

In looking for analogies to chordal graphs, it is also natural to consider *bipartite subtree graphs* (and *subtree digraphs*) in which all the sets $S_x, T_y$ (or $S_v, T_v$) are subtrees of a tree $T$. But then, as observed in [Harary, Kabell, & McMorris, 1982], every bipartite graph $G$ is a bipartite subtree graph and, as observed in [Sen, Das, & West, 1989], every digraph $D$ is a subtree digraph. For the latter, $T$ can be taken to be the bipartite graph $K_{1,|V(D)|}$ having $V(T) = V(D) \cup \{x\}$ where $x \notin V(D)$ and $E(T) = \{vx : v \in V(D)\}$; then set $S_v = \{v\}$ and $T_v = N^-(v) \cup \{x\} = \{u : uv \in A(D)\} \cup \{x\}$ for all $v \in V(D)$.

Various other notions of "bipartite chordal graphs," defined by means other than intersection, have been proposed in [Golumbic & Goss, 1978] (or section 12.4 of [Golumbic, 1980]), [McKee, 1987], and [Brandstädt, 1991]. Golumbic introduced the now-standard notion of *chordal bipartite graph*, a bipartite graph in which every cycle of length greater than four has a chord, that is discussed in detail in section 7.3. Other bipartite analogues of intersection graphs are considered in [Frost, Jacobson, Kabell, & McMorris, 1990] and [Müller, 1997].

[Harary, Kabell, & McMorris, 1990] introduces a different sort of *intersection acyclic digraph*. For instance, a digraph $D$ is an *interval acyclic digraph* if each $v \in V(D)$ can be assigned an interval $S_v$ of the real line such that the $S_v$'s all have distinct left endpoints and $uw \in A(D)$ if and only if $S_u \cap S_w \neq \emptyset$ *and* the left endpoint of $S_u$ is less than the left endpoint of $S_w$. ([McMorris & Mulder, 1996] corrects the forbidden induced subgraph characterization of interval acyclic digraph stated in the earlier paper.) [Harary, Kabell, & McMorris, 1992] extends interval acyclic digraphs to *subtree acyclic digraphs*, with vertices corresponding to subtrees of a rooted tree and with the role of left endpoints of intervals now played by vertices of the subtrees that are closest to the tree's root; the subtree acyclic digraphs constitute a proper subset of the acyclic digraphs.

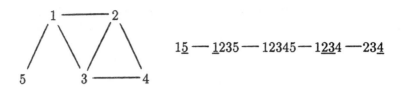

Figure 7.7: *An interval catch graph and an interval representation (with base point $b_i$ the vertex in which $i$ is underlined).*

**Theorem 7.16 (Harary, Kabell, & McMorris)** *An acyclic digraph is a subtree acyclic digraph if and only if it contains no induced subdigraph isomorphic to* •——→•←——•.

Compare this with Theorem 2.5, which essentially says that a graph is chordal if and only if it has an acyclic orientation that contains no induced subdigraph isomorphic to the digraph in Theorem 7.16; see [Rose, 1970]. [McMorris & Mulder, 1996] also considers *subpath acyclic digraphs* (using subpaths of a tree, analogous to the path graphs in section 7.1).

Related concepts were introduced (using different terminology than we use) in [Roberts, 1969a, 1971] and [Maehara, 1984a]. Given a set $S$, call a distinguished element $b \in S$ a *base point* of $S$. Given a family $\mathcal{F} = \{(S_1, b_1), \ldots, (S_n, b_n)\}$ of sets with base points ("pointed sets"), the *catch digraph* of $\mathcal{F}$ is the digraph $D$ with $V(D) = \{1, \ldots, n\}$ and $ij \in A(G)$ if and only if $i \neq j$ and $b_j \in S_i$. The *catch graph* of $\mathcal{F}$ is the graph $G$ with $V(G) = \{1, \ldots, n\}$ and $ij \in E(G)$ if and only if $i \neq j$ and either $b_j \in S_i$ or $b_i \in S_j$.

[Roberts, 1969a, 1971] focus on *interval catch graphs*, in which the $S_i$'s are intervals of a line and the base points are chosen so that $b_j \in S_i \Leftrightarrow b_i \in S_j$. Figure 7.7 shows an interval catch graph with corresponding intervals (shown as subpaths of a path as in Chapter 3) and base points (shown by underlining); for instance, if the path shown there is labeled, left to right, as $P : a, b, c, d, e$, then $S_1 = \{a, b, c, d\}$, $b_1 = b$ and $S_5 = \{a, b, c\}$, $b_5 = a$. Roberts showed that $S_i$'s and $b_i$'s can always be taken so that $b_i$ is the midpoint $S_i$ and all the $S_i$'s have the same length; moreover, as was his original motivation, interval catch graphs are precisely the graphs representable by just noticeable differences as in the application to psychology in subsection 3.4.2.

**Theorem 7.17 (Roberts)** *Interval catch graphs are precisely the proper interval graphs.*

[McKee, 1992] generalizes some of Roberts's work to catch graphs of subtrees of trees. The following, from [Ogden & Roberts, 1970], contrasts to what happens for intervals (convex subsets of a line).

**Theorem 7.18 (Ogden & Roberts)** *Every graph is the catch graph of convex subsets of a plane.*

[Maehara, 1984a] focuses on catch digraphs of $n$-dimensional spheres and boxes when the base point is the midpoint. [Sen, Das, Roy, & West, 1989] and [Prisner, 1989] study catch digraphs of intervals. [Brauner, Brualdi, & Sneyd, 1995] studies *pseudo-interval graphs*—the underlying graphs of interval catch digraphs.

## 7.3 Chordal Bipartite and Weakly Chordal Graphs

*Chordal bipartite graphs* were introduced in [Golumbic & Goss, 1978] as the bipartite graphs in which every cycle of length greater than four has a chord, equivalently, the graphs in which every induced cycle is a $C_4$. (Warning: As $C_4$ itself shows, chordal bipartite graphs need not be chordal, much as complete bipartite graphs need not be complete.)

The original motivation for chordal bipartite graphs came from applications to nonsymmetric matrices. These applications, somewhat paralleling those presented in section 2.4.3, are described in [Golumbic & Goss, 1978], [Golumbic, 1980], and [Bakonyi & Bono, 1997] (to gaussian elimination in sparse matrices); in [Hoffman, Kolen, & Sakarovitch, 1985] (to integer programming); and in [Johnson & Whitney, 1991] and [Johnson & Miller, 1997] (to matrix analysis).

Much of the literature on chordal bipartite graphs involves analogies with chordal graphs, analogies that are often edge based because of the matrix applications.

For instance, [Golumbic & Goss, 1978] defines an edge $vw \in E(G)$ to be a *bisimplicial edge* if $N(v) \cup N(w)$ induces a complete bipartite subgraph of $G$. An ordering $\langle e_1, \ldots, e_m \rangle$ of all the edges of $G$ is a *perfect edge elimination ordering* of $G$ if, for each $i \in \{1, \ldots, m\}$, $e_i$ is a bisimplicial edge of the spanning subgraph of $G$ having edge set $e_i, \ldots, e_n$. For instance, Figure 7.8 shows a chordal bipartite graph that has a perfect edge elimination ordering beginning $\langle 1b, 2b, 2c, 3c, \ldots \rangle$, with the remaining edges taken in any order. Perfect edge elimination orderings were introduced (with different names) in [Brandstädt, 1993] and [Bakonyi & Bono, 1997]. [Müller, 1997] and [Kloks &

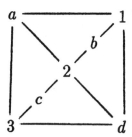

Figure 7.8: *A chordal bipartite graph.*

Kratsch, 1995] discuss algorithmic aspects of perfect edge elimination orderings and recognition of chordal bipartite graphs. ([Golumbic & Goss, 1978]
introduces a related but different notion of "perfect elimination scheme" that
is relevant to the matrix applications but for which the following theorem
fails.)

**Theorem 7.19 (Brandstädt and Bakonyi & Bono)** *A bipartite*
*graph is chordal bipartite if and only if it has a perfect edge elimination or-*
*dering.*

The following analogy of Dirac's characterization of chordal graphs is
from [Golumbic & Goss, 1978]; see also [Golumbic, 1978b]. (Warning: The
only if direction of Theorem 4 of the latter paper fails—see [Golumbic,
1980].) A set $S \subseteq V(G)$ is a *minimal edge separator* of $G$ whenever there
exist $e, f \in E(G)$ that are in different components in the subgraph induced
by $V(G) \setminus S$, and no proper subset of $S$ has this same property. For instance,
$S = \{a, d, 2\}$ is a minimal edge separator in the graph in Figure 7.8. An
independent set of vertices is said to induce a "complete bipartite subgraph"
of a bipartite graph $G$ if and only if every two of its vertices are an even
distance apart in $G$; i.e., all its vertices are of the same "color."

**Theorem 7.20 (Golumbic & Goss)** *A bipartite graph $G$ is chordal*
*bipartite if and only if every minimal edge separator induces a complete*
*bipartite subgraph of $G$.*

Recall from Theorem 2.5 that a graph is chordal if and only if its vertices
can be eliminated one at a time, where each eliminated vertex is simplicial—
which can be thought of as meaning that the vertex is not the center vertex
of an induced path of length two—in the subgraph induced by the remaining
vertices.

**Theorem 7.21 (Hammer, Maffray, & Preissmann)** *A graph is chordal bipartite if and only if its vertices can be eliminated one at a time, where each eliminated vertex is not the center vertex of an induced path of length four in the subgraph induced by the remaining vertices.*

[Hammer, Maffray, & Preissmann, 1989] also shows that if $v$ is a vertex of a chordal bipartite graph $G$ that is not the center vertex of an induced path of length four and if $w$ is a neighbor of $v$ of smallest degree, then $vw$ is a bisimplicial edge of $G$; thus a vertex elimination ordering as described in Theorem 7.21 also determines a perfect edge elimination ordering.

[Brouwer, Duchet, & Schrijver, 1983] contains the following theorem, which should be compared with Theorem 7.67. (Theorem 7.70 and the paragraph following it contain related characterizations of chordal bipartite graphs.)

**Theorem 7.22 (Brouwer, Duchet, & Schrijver)** *A graph $G$ is chordal bipartite if and only if the hypergraph $(V(G), \mathcal{E})$, with $\mathcal{E}$ the family of all open neighborhoods of $G$, is totally balanced.*

Define the *bipartite adjacency matrix* of a bipartite graph $G$ with color classes $\{a_1, \ldots, a_h\}$ and $\{b_1, \ldots, b_k\}$ to be the $h \times k$ $(0, 1)$-matrix $M = (m_{ij})$ where $m_{ij} = 1$ if and only if $a_i b_j \in E(G)$. [Hoffman, Kolen, & Sakarovitch, 1985] shows that a graph $G$ is chordal bipartite if and only if the rows and columns of its bipartite adjacency matrix can be permuted so as to contain no $\binom{1\,1}{1\,0}$ submatrix. Further refinements of this appear in [Lubiw, 1982, 1987] and [Spinrad, 1993, 1995]. This approach allows chordal bipartite graphs to be used as a tool in recognizing and studying various special kinds of graphs. For instance, [Eschen & Spinrad, 1993] use them for circular-arc graphs and [Eschen, Hayward, Spinrad, & Sritharan, to appear] use them for weakly chordal comparability graphs and weakly chordal cocomparability graphs.

Observe that the complement of a chordal graph cannot contain an induced cycle of length greater than four. This motivates the following definition from [Hayward, 1985]. A graph is *weakly chordal* (very often called *weakly triangulated*) if neither it nor its complement contains an induced cycle of length greater than four. Thus every chordal graph is weakly chordal, and so is the graph in Figure 7.8. In fact, it is easy to see the following.

**Theorem 7.23** *A graph is chordal bipartite if and only if it is both weakly chordal and bipartite.*

The following is from [Spinrad & Sritharan, 1995] and [Hayward, 1996].

**Theorem 7.24 (Spinrad & Sritharan and Hayward)** *A graph is weakly chordal if and only if its edges can be eliminated one at a time, where each eliminated edge is not the center edge of an induced path of length three in the subgraph consisting of the remaining edges.*

Algorithms on weakly chordal graphs, including recognition algorithms, make use of the following characterization from [Hayward, Hoàng, & Maffray, 1989]. [Spinrad & Sritharan, 1995] contains more details on algorithmic aspects.

**Theorem 7.25 (Hayward, Hoàng, & Maffray)** *A graph is weakly chordal if and only if every induced subgraph is either complete or contains two nonadjacent vertices such that each induced path connecting them has length two.*

## 7.4   Circle Graphs and Permutation Graphs

A graph is a *circle graph* if it is isomorphic to an intersection graph of chords of a circle. (For simplicity, all the chords can be taken to have distinct endpoints.) Circle graphs are characterized in [Even & Itai, 1971], and their early history and applications are the subject of Chapter 11 of [Golumbic, 1980], where they are introduced as "stack sorting graphs." Using stereographic projection, these are also exactly the *interval overlap graphs*—the graphs isomorphic to graphs obtained from intervals of a line with adjacency of two vertices corresponding to the intervals intersecting without either containing the other.

Other characterizations of circle graphs appear in [Fournier, 1978], [Fraysseix, 1984], [Naji, 1985] (as the consistency of a set of linear equations—[Gasse, 1997] contains a simpler proof), and [Bouchet, 1994]. [Spinrad, 1994] gives a recognition algorithm that is quadratic in the order of the graph, and [Kloks, Kratsch, & Wong, 1996] gives a cubic algorithm for the minimum fill-in problem.

For any vertex $v$ of a graph $G$, define a *local complementation* of $G$ at $v$ to be the graph obtained by replacing $N(v)$—the graph induced by the neighbors of $v$—by its complement. Two graphs are *locally equivalent* if one can be obtained from the other by a sequence of local complementations.

**Theorem 7.26 (Bouchet)** *A graph is a circle graph if and only if it contains no subgraph that is locally equivalent to one of the graphs shown in Figure 7.9.*

Figure 7.9: *Three graphs that are not circle graphs.*

If the radius of the circle is thought of as infinitely large, then the chords become intersecting or parallel lines and the associated intersection graphs are precisely the complete multipartite graphs (with parallel classes of lines corresponding to indcpcndent "parts" of the graph)—*if* we happen to be in the *euclidean* plane. In elliptic geometry there are no parallel lines, and so the intersection graphs of lines are precisely the complete graphs. While the question would seem to be much harder in hyperbolic geometry, the common "Beltrami–Klein circle model," as for instance in [Greenberg, 1980], shows that the intersection graphs of lines are again precisely the circle graphs.

[Elmallah & Stewart, 1993] defines a *k-polygon graph* to be a graph isomorphic to the intersection graph of line segments drawn between points on distinct sides of a *k*-sided polygon. The circle graphs are precisely the graphs that are *k*-polygon graphs for some *k*.

The smallest *k* for which *G* is a *k*-polygon graph is a measure of how far *G* is from being a *permutation graph*, an intersection graph of line segments drawn between two parallel lines (so a sort of "2-polygon graph"). Permutation graphs are frequently useful in specific computational problems, and so they have been looked at in a wide variety of contexts. We, however, only mention their structural properties. Chapter 7 of [Golumbic, 1980] is a standard reference, and [Pnueli, Lempel, & Even, 1971] is a key paper. Permutation graphs are special sorts of trapezoid graphs (section 7.1) and of asteroidal triple-free graphs (section 7.6).

The following motivates the name "permutation." Suppose $\pi$ is any permutation of $\{1, \ldots, n\}$ and consider the resulting list $[\pi(1), \ldots, \pi(n)]$. The *permutation graph* $G(\pi)$ has vertices $v_1, \ldots, v_n$ with an edge between $v_i$ and $v_j$ whenever $i$ and $j$ occur "out of order" in the list. In other words, $v_i v_j \in E(G(\pi))$ if and only if $i < j$ and $i$ is to the right of $j$ in the list, meaning that $\pi^{-1}(i) > \pi^{-1}(j)$. Alternatively, if you place $\langle 1, \ldots, n \rangle$ and $\langle \pi(1), \ldots, \pi(n) \rangle$ on parallel lines and draw the $n$ line segments for each $i, \pi(i)$ pair, then $G(\pi)$ is isomorphic to the intersection graph of these line

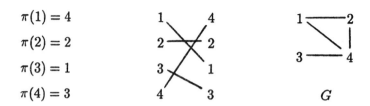

$$\begin{aligned}\pi(1) &= 4\\ \pi(2) &= 2\\ \pi(3) &= 1\\ \pi(4) &= 3\end{aligned}$$

Figure 7.10: *A permutation graph example.*

segments.

As an example, suppose $n = 4$, with $\pi(1) = 4$, $\pi(2) = 2$, $\pi(3) = 1$, and $\pi(4) = 3$. Then the list is $[4, 2, 1, 3]$ and four line segments and the permutation graph are as shown in Figure 7.10.

By looking at the permutation with reversed list, $[3, 1, 2, 4]$ in the example, it is easy to see that the complement $\overline{G}$ of a permutation graph $G$ is also a permutation graph. [Pnueli, Lempel, & Even, 1971] relates permutation graphs with containment and comparability graphs (section 7.6) using Theorem 7.36.

**Theorem 7.27 (Pnueli, Lempel, & Even)** *A graph $G$ is a permutation graph if and only if both $G$ and $\overline{G}$ are comparability graphs and so if and only if $G$ is the containment graph of intervals of a line.*

This led to considerable work on polynomial recognition algorithms for permutation graphs, culminating in a linear-time recognition algorithm in [McConnell & Spinrad 1994].

## 7.5   Clique Graphs of Chordal Graphs and Clique-Helly Graphs

The clique graphs of chordal graphs were independently studied in [Brandstädt, Dragan, Chepoi, & Voloshin, 1994] as *dually chordal graphs* (called "HT-graphs" in earlier work in, for instance, [Dragan, 1993]), in [Szwarcfiter & Bornstein, 1994] as "expanded trees," and in [Gutierrez & Oubiña, 1996] as "tree-clique graphs" (based on earlier work in [Batbedat, 1990]). We include only some of their characterizations.

**Theorem 7.28 (Szwarcfiter & Bornstein)** *A graph $G$ is the clique graph of a chordal graph if and only if $G$ has a spanning tree $T$ such that, for*

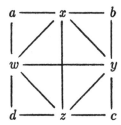 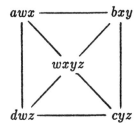

Figure 7.11: *A chordal graph $G$ and its clique graph $K(G)$.*

each $uv \in E(G)$, the vertices of the $u$-to-$v$ subpath in $T$ induce a complete subgraph in $G$.

This appears in a slightly different form in [Brandstädt, Dragan, Chepoi, & Voloshin, 1994] and [Gutierrez & Oubiña, 1996]: $G$ is the clique graph of a chordal graph if and only if $G$ has a spanning tree $T$ such that, for each maxclique $Q$ of $G$, the vertices of $Q$ induce a subtree of $T$.

Figure 7.11 shows a chordal graph $G$, which is not the clique graph of a chordal graph, and its clique graph $K(G)$, which is not a chordal graph. The spanning tree of $K(G)$ described in the preceding theorem consists of the four spokes of the wheel.

The motivation for calling these "dually chordal graphs" is that, in the terminology of section 2.3, $G$ is the clique graph of a chordal graph if and only if $G$'s clique hypergraph is a tree hypergraph, whereas $G$ is a chordal graph if and only if the dual of $G$'s clique hypergraph is a tree hypergraph.

**Theorem 7.29 (Brandstädt, Dragan, Chepoi, & Voloshin)** *A graph $G$ is the clique graph of a chordal graph if and only if $V(G)$ can be ordered $\langle v_1, \ldots, v_n \rangle$ such that, for each $v_i$, there is a $v_j$ with $j \geq i$ such that, relative to the subgraph of $G$ induced by $v_i, \ldots, v_n$, $N[w] \subseteq N[v_j]$ for each $w \in N[v_i]$.*

[Brandstädt, Chepoi, & Dragan, 1995] gives a recognition algorithm that finds such vertex orderings. [Brandstädt, Dragan, Chepoi & Voloshin, 1994] shows that if $G$ is the clique graph of a chordal graph, then so is every power of $G$. That paper also contains the following connection with strongly chordal graphs (section 7.12).

**Theorem 7.30 (Brandstädt, Dragan, Chepoi, & Voloshin)** *A graph $G$ is strongly chordal if and only if every induced subgraph of $G$ is the clique graph of a chordal graph.*

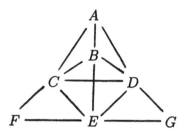

Figure 7.12: *A clique graph that is not a clique-Helly graph.*

[Bandelt & Prisner, 1991] shows that $K(G)$ is chordal whenever $G$ is the clique graph of a chordal graph.

A graph $G$ is a *clique-Helly graph* if the family of all the maxcliques of $G$ satisfies the Helly condition—if the members of any family of maxcliques of $G$ intersect pairwise, then they all have a common element. Clique-Helly graphs were first studied in [Hamelink, 1968], showing that clique-Helly graphs are clique graphs. Figure 7.12 shows a clique graph (Exercise 1.19) that is not clique-Helly. [Escalante, 1973] strengthens Hamelink's result.

**Theorem 7.31 (Escalante)** *A graph $G$ is a clique-Helly graph if and only if $G = K(H)$ where $H$ is another clique-Helly graph.*

For every triangle $uvw$ of a graph $G$, let $G_{uvw}$ denote the subgraph of $G$ induced by those vertices that are adjacent to at least two of $u, v, w$. [Szwarcfiter, 1997] proves that a graph $G$ is clique-Helly if and only if every such $G_{uvw}$ contains a vertex adjacent to all the other vertices of $G_{uvw}$. This also gives an efficient recognition algorithm.

Let $K^{n+1}(G) = K(K^n(G))$, where $K^1(G) = K(G)$. The following is from [Brandstädt, Dragan, Chepoi, & Voloshin, 1994].

**Theorem 7.32 (Brandstädt, Dragan, Chepoi, & Voloshin)** *A graph $G$ is both chordal and clique-Helly if and only if $G = K^2(H)$ where $H$ is a chordal graph (and so if and only if $G = K(H)$ where $H$ is a clique graph of a chordal graph).*

[Escalante, 1973] and [Bandelt & Prisner, 1991] show that every clique-Helly graph $G$ has parameters $p \leq 2$ and $n$ such that $K^{n+p}(G) \cong K^n(G)$. This leads into many questions involving iterating the clique graph operator; see, for instance, [Bandelt & Prisner, 1991], [Prisner, 1995], and [Bornstein & Szwarcfiter, 1995].

[Prisner, 1993] and [Wallis & Zhang, 1990] study *hereditary clique-Helly graphs*—graphs for which every induced subgraph is a clique-Helly graph. A family $\mathcal{F} = \{S_1, \ldots, S_k\}$ of subsets of a set $S$ is said to satisfy the *strong Helly condition* if, for every subfamily $\mathcal{F}' \subseteq \mathcal{F}$,

$$|\bigcap\{S_i : S_i \in \mathcal{F}'\}| = \min\{|S_i \cap S_j| : S_i, S_j \in \mathcal{F}' \text{ and } i \neq j\}.$$

By induction, this is equivalent to, for every three members $S_i, S_j, S_k \in \mathcal{F}$,

$$|S_i \cap S_j \cap S_k| = \min\{|S_i \cap S_j|, |S_i \cap S_k|, |S_j \cap S_k|\}.$$

It can be shown that $\mathcal{F}$ satisfies the strong Helly condition if and only if the hypergraph $(V(G), \mathcal{F})$ is a strong Helly hypergraph (section 2.3).

**Theorem 7.33 (Prisner)** *A graph $G$ is a hereditary clique-Helly graph if and only if the family of all the maxcliques of $G$ satisfies the strong Helly condition.*

[Wallis & Zhang, 1990] defines a graph $G$ to be *irreducible* if each max-clique of $G$ contains an edge that is in no other maxclique. Based on that work, Prisner gives a forbidden induced subgraph characterization of hereditary clique-Helly graphs and shows that a graph is a hereditary clique-Helly graph if and only if each of its induced subgraphs is irreducible.

**Theorem 7.34 (Prisner)** *A graph $G$ is a hereditary clique-Helly graph if and only if $G = K(H)$ where $H$ is another hereditary clique-Helly graph.*

[McKee, 1994] defines a notion of an "absolutely clique-Helly pseudo-graph" that characterizes those pseudographs from section 6.2 that have pseudo duals.

# 7.6   Containment, Comparability, Cocomparability, and Asteroidal Triple-Free Graphs

A graph $G$ is a *containment graph* of some family $\mathcal{F} = \{S_1, \ldots, S_n\}$ of nonempty sets if $V(G) = \mathcal{F}$ and $S_i S_j \in E(G)$ if and only if one of $S_i$ and $S_j$ is properly contained in the other. Recall that $(X, <)$ is a *partially ordered set* (or *poset*) if $<$ is an irreflexive, transitive binary relation on the nonempty set $X$. [Golumbic & Scheinerman, 1989] observes that a graph $G$ is a containment graph if and only if it is a *comparability graph* of a poset $(X, <)$, where this means that $V(G) = X$ and $uv \in E(G)$ if and only if either

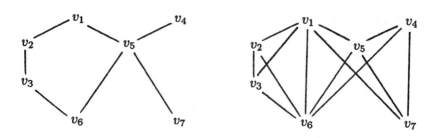

Figure 7.13: *A poset and its comparability graph.*

$u < v$ or $v < u$. Alternatively, a graph $G$ is a comparability graph if and only if $E(G)$ can be *transitively oriented*—meaning that an orientation $D$ of $G$ has arc set $A(D)$ such that $uv, vw \in A(D)$ implies $uw \in A(D)$. Figure 7.13 shows a poset on the left, with $v_i < v_j$ whenever $v_i$ is below $v_j$ on a path, and its comparability graph $G$ on the right; $G$ is also the containment graph of the "downsets" $S_i = \{j : v_j \leq v_i\}$ and can be transitively oriented by directing all arcs downward. Chapter 8 of [Golumbic, 1980] is a standard reference on comparability graphs, and [Hell & Huang, 1995] gives an up-to-date description of algorithms.

The following two results, from [Golumbic & Scheinerman, 1989] and [Dushnik & Miller, 1941], respectively, show that containment graphs behave very differently from intersection graphs with regard to the topics in Chapters 2 and 3.

**Theorem 7.35 (Golumbic & Scheinerman)** *Every containment graph is the containment graph of a family of subtrees of a tree.*

**Theorem 7.36 (Dushnik & Miller)** *A graph $G$ is the containment graph of a family of intervals of the real line if and only if both $G$ and its complement $\overline{G}$ are containment graphs.*

As a corollary of this and Theorem 7.27, $G$ is the containment graph of a family of intervals of the real line if and only if $G$ is a permutation graph (section 7.4). Dushnik & Miller show that these containment graphs are also precisely the comparability graphs of posets of "dimension two."

Paralleling section 1.2, [Golumbic & Scheinerman, 1989] characterizes *containment classes*, sets of graphs that are exactly those isomorphic to containment graphs of arbitrary families $\mathcal{F}$ of members of some set $\Sigma$ of sets (for instance, the set of all subtrees of a tree or of all intervals of the real line). Define a set $\mathcal{G}$ of graphs to be *closed under vertex multiplication* if

$G' \in \mathcal{G}$ whenever $G'$ results from $G \in \mathcal{G}$ by repeatedly replacing an existing vertex $v$ by a pair $v', v''$ of new *non*adjacent vertices, each having the same pre-existing neighbors as $v$ did.

**Theorem 7.37 (Golumbic & Scheinerman)** *A set $\mathcal{G}$ of comparability graphs is a containment class if and only if all three of the following conditions are satisfied:*
(1) *$\mathcal{G}$ is closed under induced subgraphs;*
(2) *$\mathcal{G}$ is closed under vertex multiplication;*
(3) *$\mathcal{G}$ has a composition series.*
*Moreover, if repeated members of $\Sigma$ are not allowed in the $\mathcal{F}$'s, then conditions (1) and (3) are necessary and sufficient.*

[Golumbic & Scheinerman, 1989] also characterizes related kinds of graphs and classes based on overlap and disjointedness, instead of intersection or comparability.

[Sen, Sanyal, & West, 1995] considers a directed version of containment graphs. [Ma & Spinrad, 1991] studies chordal comparability graphs and [Eschen, Hayward, Spinrad, & Sritharan, to appear] studies weakly chordal comparability graphs.

[Golumbic & Scheinerman, 1989] also contains results on $d$-box containment, paralleling section 7.1. [McKee & McMorris, 1992] discusses comparability multigraphs, paralleling section 6.2. [McKee, 1995b] introduces "connection graphs," generalizing both intersection and containment graphs (and also catch graphs as in section 7.2), emphasizing the appropriate analogues of edge clique covers and characterizing "connection classes."

A graph $G$ is a *cocomparability graph* if its complement $\overline{G}$ is a comparability graph (or, equivalently, a containment graph). Theorem 3.5 says that a graph is an interval graph if and only if it is a cocomparability graph that contains no induced cycle $C_4$, and Theorem 7.27 says that a graph is a permutation graph if and only if it is both a comparability graph and a cocomparability graph.

[Corneil & Kamula, 1987] shows that every trapezoid graph (section 7.1) is a cocomparability graph, and so that is also true of interval graphs and permutation graphs. Conversely, [Felsner, 1993] shows that every cocomparability graph that is also a min-tolerance interval graph must be a trapezoid graph. [Langley, 1994] shows that every bipartite cocomparability graph is a min-tolerance interval graph.

The following is from [Golumbic, Rotem, & Urrutia, 1983].

**Theorem 7.38 (Golumbic, Rotem, & Urrutia)** *A graph is a cocomparability graph if and only if it is the intersection graph of continuous functions $f : [0, 1] \to \mathbb{R}$, each viewed as a set of points in $\mathbb{R}^2$.*

Just as interval, permutation, and trapezoid graphs have linear structures, so do cocomparability graphs: a graph $G$ is a cocomparability graph if and only if its vertices can be arranged in a path such that $uv \in E(G)$ if and only if, for each vertex $x$ in between $u$ and $v$ on the path, either $ux \in E(G)$ or $vx \in E(G)$.

Interval, permutation, trapezoid, and cocomparability graphs are all special cases of *asteroidal triple-free graphs*—graphs that contain no asteroidal triples (section 3.1). [Corneil, Olariu, & Stewart, 1997] is an excellent survey and synthesis of all aspects of these only recently studied graphs (including, for instance, a cubic recognition algorithm). The following two theorems are from that paper, with one direction of the second from [Möhring, 1996]. (Notice the ramifications for the minimum fill-in problem.) Recall that $S \subseteq V(G)$ *dominates* $G$ if every $w \in V(G)$ is either in $S$ or is adjacent to a vertex in $S$.

**Theorem 7.39 (Corneil, Olariu, & Stewart)** *Every connected asteroidal triple-free graph $G$ contains a pair $u, v$ of vertices such that every $u, v$-path dominates $G$ (and $u, v$ can be found such that their distance in $G$ equals the diameter of $G$).*

**Theorem 7.40 (Möhring and Corneil, Olariu, & Stewart)** *A graph $G$ is asteroidal triple-free if and only if adding a minimal set of new edges to $G$ so as to create a chordal graph always creates an interval graph.*

## 7.7   Infinite Intersection Graphs

Although we have been making the common graph-theoretic restriction to finite vertex sets throughout the rest of this monograph, much work on intersection graphs has involved infinite graphs. [Diestel, 1990] is a recent, exhaustive survey of many aspects of this work.

Interestingly, the earliest paper on chordal graphs, [Hajnal & Surányi, 1958], was most definitely interested in the infinite case in connection with the "Souslin hypothesis" (that the real line can be characterized as a dense linear order without endpoints, complete under the formation of sups and infs, such that every collection of pairwise disjoint open intervals is countable). [Wolk, 1962] also introduced $P_4$-free chordal graphs (section 7.9) with the Souslin hypothesis as motivation.

Define an *infinite subtree graph* to be the intersection graph of an infinite family of subtrees of an infinite tree and an *infinite chordal graph* to be a graph that contains no induced cycles larger than triangles. While every infinite subtree graph is an infinite chordal graph, the converse was disproved in [Halin, 1984], even for graphs with just a countably infinite vertex set. [Diestel, 1988] gives the following example of a countable chordal graph $H_1$ that is not a subtree graph: take $V(H_1) = \{x_1, x_2, \ldots; s_1, s_2, \ldots; q\}$ and $E(H_1) = \{x_i x_{i+1} : i \geq 1\} \cup \{x_i s_j : 1 \leq i \leq j\} \cup \{s_i s_j : i, j \geq 1\} \cup \{s_i q : i \geq 1\}$. Another example is given by $H_2$, which is obtained from $H_1$ by including all edges $x_i x_j$ where $i, j \geq 1$. Diestel then proves that a countable graph is a subtree graph if and only if it is chordal and contains neither $H_1$ nor $H_2$ as a "simplicial minor."

Note that each of Diestel's two problematic subgraphs contains an infinite complete subgraph. [Halin, 1984] characterizes the infinite subtree graphs in terms of a suitable version of perfect elimination orderings from which the following is a corollary.

**Theorem 7.41 (Halin)** *A graph with no infinite complete subgraphs is an infinite subtree graph if and only if it is an infinite chordal graph.*

[Halin, 1982] also considers *infinite interval graphs*, meaning the intersection graph of an infinite number of intervals of the real line.

**Theorem 7.42 (Halin)** *A graph is an infinite interval graph if and only if every finite induced subgraph is an interval graph and is equivalent to every three maxcliques having one that separates the other two.*

## 7.8 Miscellaneous Topics

**Completion Sequences.** A *completion sequence* for a graph $G$ within a class of graphs is a sequence of edges of the complement $\overline{G}$ such that, when these edges are inserted one at a time, each of the resulting graphs from $G$ up to $K_{|V(G)|}$ is also in the class. This notion was introduced in [Grone, Johnson, Sá, & Wolkowicz, 1984], where it is shown that the class of chordal graphs allows such completion sequences. [Rasmussen, 1994] shows that the classes of chordal, interval, proper interval, split, circular-arc, proper circular-arc, comparability, and permutation graphs allow completion sequences. [Odom & Rasmussen, 1995] adds strongly chordal graphs to this list and emphasizes polynomial algorithms for finding completion sequences.

[Bakonyi & Bono, 1997] contains the corresponding result for chordal bipartite graphs. See [Spinrad & Sritharan, 1995] for a related approach to weakly chordal graphs.

**Dot Product Representations.** [Fiduccia, Scheinerman, Trenk, & Zito, 1998] defines the *dot product graph* of a family $\mathcal{F}$ of $k$-tuples of reals to have vertex set $\mathcal{F}$ with vectors $\vec{v}, \vec{w} \in \mathcal{F}$ adjacent if and only if $\vec{v} \cdot \vec{w} \geq 1$. This generalizes the intersection graph $\Omega(\{S_1, \ldots, S_n\})$ in that each set $S_i$ corresponds to a characteristic vector $\vec{v}_i$ in $\{0, 1\}^{|\cup_i S_i|}$ and $S_i \cap S_j \neq \emptyset$ if and only if $\vec{v}_i \cdot \vec{v}_j \geq 1$.

Fiduccia, Scheinerman, Trenk, & Zito define the *dot product dimension* of $G$ to be the minimum $k$ such that $G$ is the dot product graph of a set of $k$-vectors. The dot product dimension of a graph is less than or equal to the intersection number of the graph. Among a wealth of results on dot product graphs (and their generalizations!), they show that every interval graph has dot product dimension at most two and every chordal graph $G$ has dot product dimension at most one plus the order of the largest maxclique in $G$.

**Fuzzy Intersection Graphs.** A *fuzzy set* is a set in which each potential element is in the set with a particular value ("degree of membership") between 0 and 1. Fuzzy set theory (along with fuzzy logic and the like) is currently popular among certain mathematicians and computer scientists. A "fuzzy graph" is a graph in which each pair of vertices is joined by a "fuzzy edge" with a value between 0 and 1. *Fuzzy intersection graphs*—defined in terms of "fuzzy intersection" of fuzzy sets—are discussed in [McAllister, 1988].

See [Craine, 1994] for another notion of fuzzy intersection graphs with an analogue of Marczewski's theorem (Theorem 1.1) and a notion of "fuzzy interval graph" that has an analogue of the Gilmore–Hoffman characterization (Theorem 3.5) but not of the Fulkerson–Gross characterization (Corollary 3.2).

**Intersection Graphs from Designs.** Design theory is one of the central areas of combinatorics and has many applications. A *design* on a set is a collection of subsets, called *blocks*, such that every pair of elements of the underlying set is contained in a fixed number of blocks. The *block-intersection graph of a design* (not to be confused with a graph-theoretic "block graph") is the intersection graph of its blocks. See [Alspach & Hare, 1991] and its references for results on (and extending) the hamiltonicity of the block-intersection graph, and [Hare & McCuaig, 1993] for discussion of questions of connectivity.

**Intersection Graphs of Algebraic Structures.** [Csákány & Pollák, 1969] defined the *graph of subgroups of a group* $G$ to be the intersection graph of every $H \setminus \{e\}$, where $H$ is a proper, nontrivial subgroup of $G$ and $e$ is the identity element of $G$. The study of the interplay of the structures of the group and its graph was continued in [Zelinka, 1975a]. All sorts of algebraic structures can be similarly treated, with [Bosák, 1975] considering the graph of subalgebras of an algebra.

Some of the earliest work involved the *graph $G(S)$ of subsemigroups of a semigroup of a group $S$*, the intersection graph of all the proper subsemigroups of $S$. This began with [Bosák, 1964] and continued over the next decade, considering such things as the connectivity, diameter, and girth of $G(S)$; see [Shevrin & Ovsyannikov, 1983] for references. The topic was resurrected in [Luedeman & McMorris, 1986], characterizing when $G(S)$ is a tree, and in [Ackerman, McMorris, & Seif, 1993], focusing on the question of when $G(S)$ is chordal. [Luedeman & McMorris, 1986] also studies the intersection graphs of right ideals, [Luedeman, 1987] the intersection graphs of quasi-ideals and bi-ideals of semigroups, and [Pondělíček, to appear] the intersection graphs of semigroups in which every element is idempotent.

It is easy to see that graphs $G \cong G(S)$ for a semigroup $S$ are upper bound graphs, but a complete characterization of such graphs remains an interesting open problem.

**Intersection Graphs of Graphs.** [Zelinka, 1975b] defines the *intersection graph of a graph* $G$ to be the intersection graph of the edge sets of all the proper induced subgraphs of $G$. The paper contains results and examples concerning the conjecture that every graph of order at least four is uniquely determined by its intersection graph.

**Partition Graphs.** Partition graphs (not to be confused with the "partition intersection graphs" described in section 2.4) were introduced in [De-Temple, Robertson, & Harary, 1984]. A *partition graph* is an intersection graph $G$ of a family of subsets of a set $S$ such that the vertices in every maximal independent subset of $V(G)$ correspond to a partition of $S$. The graph in Figure 7.14 is an example of a partition graph on the set $S = \{1, 2, 3, 4, 5\}$, as is witnessed by the set-labeled intersection representation shown there: the maximum independent subsets of $V(G)$ are $\{a, e\}$, $\{b, f\}$, $\{c, d\}$, $\{a, d, f\}$, and each corresponds to a partition of $S$.

[McAvaney, Robertson, & DeTemple, 1993] characterizes partition graphs

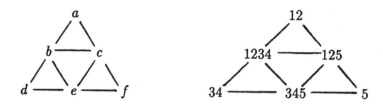

Figure 7.14: *A partition graph $G$ on the set $\{1, 2, 3, 4, 5\}$.*

by the existence of an edge clique cover $\mathcal{Q}$ such that each $Q$ is a maxclique that has a nonempty intersection with every maximal independent set of vertices. [DeTemple, Dineen, Robertson, & McAvaney, 1993] contains related details.

**Random Intersection Graphs.** The various models for *random graphs* provide powerful tools for understanding many graph-theoretic concepts, including intersection graphs. [Janson & Kratochvíl, 1992] has a broad discussion; see also [Maehara, 1991].

Random interval graphs are discussed in [Scheinerman, 1988b, 1990b] and applied to queuing theory in [Nawijn, 1991]. See [Scheinerman, 1990a] for connections with interval numbers as in section 7.1. [McMorris & Scheinerman, 1991] discusses random chordal graphs, and [Maehara, 1990] discusses random circular-arc graphs (as in section 7.1).

## 7.9  $P_4$-**Free Chordal Graphs and Cographs**

The $P_4$-*free chordal graphs* are easily seen to be the same as the $C_4$, $P_4$-free graphs: the graphs that have no induced 4-vertex cycle or path. These graphs form one of several well-studied forbidden-subgraph subclasses of chordal graphs; we include a section on this particular one because of its many relationships to other concepts we have considered. These graphs first appeared in [Wolk, 1962, 1965].

**Theorem 7.43 (Wolk)** *A graph is a $P_4$-free chordal graph if and only if it is the comparability graph of a tree poset.*

More directly relevant to intersection graphs, [Skrien, 1982] defines a graph to be a *nested interval graph* if it is the intersection graph of a family of *nested intervals* of the real line, meaning that two intervals in the family have a nonempty intersection only when one of them is contained in the other. Equivalently, a graph is a nested interval graph if and only if it is the containment graph (section 7.6) of nested intervals.

**Theorem 7.44 (Skrien)** *A graph is a $P_4$-free chordal graph if and only if it is a nested interval graph.*

A graph $G$ is *perfect* if, for every induced subgraph $G'$ of $G$, the cardinality of the largest independent set in $G'$ equals the minimum number of maxcliques needed to cover $V(G')$. (Perfect graphs, introduced by Berge in the early 1960s, have an immense literature that includes [Golumbic, 1980], the classic textbook on intersection graph theory.) [Golumbic, 1978a] defines a graph $G$ to be *trivially perfect* if, for every induced subgraph $G'$ of $G$, the cardinality of the largest independent set in $G'$ equals the number of maxcliques of $G'$.

**Theorem 7.45 (Golumbic)** *A graph is a $P_4$-free chordal graph if and only if it is trivially perfect.*

A set $S \subseteq V(G)$ *dominates* a graph $G$ if every vertex of $G$ is either in $S$ or has a neighbor in $S$. The *domination number* of $G$, denoted $\gamma(G)$, is the smallest cardinality of a set $S$ that dominates $G$. For any complete subgraph $Q$ of $G$, define $N(Q)$, the *common neighborhood* of $Q$, to be $\bigcap \{N(v) : v \in Q\}$. The following theorems are from [McKee, 1990c, to appear(a)], respectively; there is related material in [Kelleher & Cozzens, 1990].

**Theorem 7.46 (McKee)** *A graph $G$ is a $P_4$-free chordal graph if and only if $V(G)$ can be ordered $\langle v_1, \ldots, v_n \rangle$ where each $v_i$ dominates its component in the subgraph of $G$ induced by $v_i, \ldots, v_n$.*

**Theorem 7.47 (McKee)** *For every graph $G$,*

$$\sum_Q [1 - \gamma(N(Q))] \leq \gamma(G),$$

*where the sum is taken over all nonempty complete subgraphs $Q$ of $G$, with equality holding if and only if $G$ is a $P_4$-free chordal graph. (The inequality also holds when the $\gamma$ parameter is replaced with the number of components, with equality then holding if and only if $G$ is chordal.)*

Define a graph to be a *hereditary upper bound graph* if every induced subgraph is an upper bound graph. [Myers, 1982] shows that a graph is a $P_4$-free chordal graph if and only if it is a hereditary upper bound graph. [Ma, Wallis, & Wu, 1989] characterize $P_4$-free chordal graphs as *quasi-threshold graphs*, a weakening of the notion of threshold graphs from Chapter 5; [Yan, Chen, & Chang, 1996] discusses quasi-threshold graphs further. $P_4$-free chordal graphs also show up in [Peyton, Pothen, & Yuan, 1995] in connection with sparse matrix computations.

[Skrien, 1982] refers (equivalently) to the $P_4$-free chordal graphs as being the $P_4$-free *interval* graphs. Replacing $P_4$ with the 5-vertex tree F having degrees 1, 1, 1, 2, and 3 (like the letter F), [McKee, 1998] shows that F-free interval graphs are those for which clique path representations are produced by a simple-minded greedy path algorithm obtained by modifying Kruskal's algorithm to repeatedly choose an edge of largest weight that does not form either a cycle or a vertex of degree three with previously chosen edges. This gives an analogue of the chordal graph greedy tree algorithm of Theorem 2.3: The $P_4$-free interval graphs (equivalently, the nested interval graphs) are precisely the graphs for which the simple-minded greedy path algorithm produces a nested interval representation.

Those $P_4$-free graphs that are not necessarily chordal have also been independently investigated and are also known by many different names, the most frequent being *complement reducible graphs* or, more often, *cographs*. [Corneil, Lerchs, & Stewart Burlingham, 1981] is the key paper, organizing many people's work from the 1970s, with an assortment of characterizations and applications. [Chaiken, Murray, & Rosenthal, 1989] contains even more, featuring an application to automated theorem proving. Cographs form a subclass of the permutation graphs (section 7.4). The "complement reducible" name comes from the following.

**Theorem 7.48 (Corneil, Lerchs, & Stewart Burlingham)** *A graph is a cograph if and only if it can be reduced to an edgeless graph by repeatedly taking complements within components. In other words, if $G^{(0)} = G$ and $G^{(i+1)}$ is the union of the complements of all the components of $G^{(i)}$, then $G$ is a cograph if and only if these $G^{(i)}$ all exist and become edgeless for sufficiently large $i$.*

Figure 7.15 shows an example of a cograph $G$ and $G^{(1)}$ and $G^{(2)}$; the $G^{(i)}$ with $i \geq 3$ are all edgeless. Trying the same process on $P_4$ clearly never even begins to lead to an edgeless graph.

Paralleling Theorem 7.43, a graph is a cograph if and only if it is the comparability graph of a series-parallel poset. Paralleling Theorem 7.45,

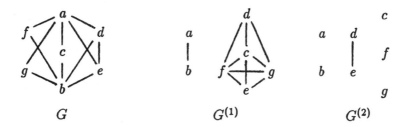

Figure 7.15: *A cograph $G$ with its nonedgeless subsidiary cographs $G^{(i)}$.*

$G$ is a cograph if and only if, in every induced subgraph $G'$ of $G$, every maxclique and every maximal independent set have exactly one vertex in common. Other simple characterizations include the following:

- In every nontrivial (meaning $|V(G')| > 1$) induced subgraph $G'$ of $G$, there are vertices $u, w$ such that $N(u) \setminus \{u, w\} = N(w) \setminus \{u, w\}$ in $G'$.

- For every nontrivial induced subgraph $G'$ of $G$, either $G'$ or its complement $\overline{G'}$ is not connected.

- Every connected induced subgraph of $G$ has diameter at most two.

- $G$ can be generated from trivial graphs by a sequence of disjoint unions and joins.

[McKee, 1990c] further studies the connections among cographs, intersection graphs, and comparability graphs (as well as their multigraph analogues), including the statement of the following "odd" intersection characterization. Call a graph the *odd intersection graph* of a family $\mathcal{F} = \{S_1, \ldots, S_n\}$ of sets if it has $\mathcal{F}$ as its vertex set with $S_i$ and $S_j$ adjacent if and only if $|S_i \cap S_j|$ is odd.

**Theorem 7.49** *A graph $G$ is a cograph if and only if there exists a rooted tree $T$ with no nonroot vertex of degree two such that $G$ is isomorphic to the odd intersection graph of all the root-to-leaf paths of $T$.*

These trees are essentially the *cotrees* in [Corneil, Lerchs, & Stewart Burlingham, 1981]. For instance, a cotree representation for the cograph in Figure 7.15 is shown in Figure 7.16; notice how its subtrees correspond to the components of the $G^{(i)}$'s. [Corneil, Perl, & Stewart, 1985] contains more information on cographs, including a linear algorithm for recognizing cographs and constructing cotree representations.

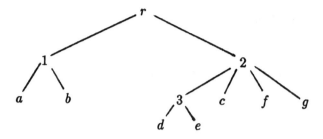

Figure 7.16: *A cotree representation for the cograph in Figure 7.15.*

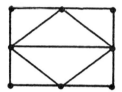

Figure 7.17: *A chordal graph whose square is not chordal.*

## 7.10   Powers of Intersection Graphs

Generalizing the notion of squared graph from section 4.1.1, the *k-power* of a graph $G$, denoted $G^k$, has the same vertices as $G$, with two vertices $u$ and $w$ adjacent if and only if $d(u, w) \leq k$, where $d(u, w)$ denotes the usual graph distance in $G$. Section 16.2 of [Prisner, 1995] discusses powers of all sorts of intersection graphs.

Many intersection classes $\mathcal{G}$ are *closed under powers*, meaning that $G \in \mathcal{G}$ implies that $G^k \in \mathcal{G}$ for all $k \geq 1$. An intersection class $\mathcal{G}$ is *strongly closed under powers* if, for every $k \geq 1$, $G^k \in \mathcal{G}$ implies that $G^{k+1} \in \mathcal{G}$; in other words, if a power of $G$ is in the class, then so are all higher powers.

[Laskar & Shier, 1980] effectively began the study of powers of chordal graphs, noting that the class of all chordal graphs is not closed under powers: the example in Figure 7.17 is a chordal graph whose square is not chordal. [Laskar & Shier, 1983] and [Wallis & Wu, 1995] both characterize when the square of a chordal graph is chordal.

**Theorem 7.50 (Wallis & Wu)** *A chordal graph $G$ has $G^2$ chordal if and only if the clique graph $K(G)$ of $G$ is chordal. For any graph $G$, $K(G)$ chordal implies $G^2$ is chordal.*

[Duchet, 1984], [Balakrishnan & Paulraja, 1983], and [Flotow, 1997] contain further results on powers of chordal graphs. Call the graph formed from a triangle by adding two nonadjacent pendant edges (like the letter A) an A-graph, and recall that $K_{1,3}$ is the upper-left graph in Figure 1.4.

**Theorem 7.51 (Duchet)** *If $G^k$ is chordal, then so is $G^{k+2}$. If $G$ and $G^2$ are both chordal, then all powers of $G$ are chordal.*

**Theorem 7.52 (Balakrishnan & Paulraja)** *If $G$ is chordal and $k$ is odd, then $G^k$ is chordal. If $G$ is chordal and $G^{2k}$ is not chordal, then none of the edges of any chordless cycle of $G^{2k}$ is an edge of $G^r$ for $r < 2k$.*

**Theorem 7.53 (Flotow)** *If $G$ contains no induced A-graph or $K_{1,3}$ or $C_n$ with $n \geq 2k + 2$, then $G^k$ is chordal. If $G$ contains no induced $K_{1,3}$ or $C_n$ with $n \geq 4$, then every power of $G$ is chordal. If $G$ contains no induced $K_{1,3}$ or $C_n$ with $n \geq 6$, then every odd power $G^k$, $k \geq 3$, of $G$ is chordal.*

[Brandstädt, Chepoi, & Dragan, 1996] shows that if $G^m$ and $G^n$ are both chordal, then they have a common perfect elimination ordering that can be found efficiently using a modified maximum cardinality search.
[Raychaudhuri, 1987], [Prisner, 1996b], and [Flotow, 1996] investigate the classes of (proper) interval graphs and circular-arc graphs.

**Theorem 7.54 (Raychaudhuri)** *The intersection classes of interval graphs, asteroidal triple-free graphs, and proper interval graphs are strongly closed under powers.*

**Theorem 7.55 (Prisner)** *The intersection class of proper circular-arc graphs is strongly closed under powers.*

**Theorem 7.56 (Flotow)** *The intersection class of circular-arc graphs is closed under powers, and for all $k \geq 2$, $G^k$ a circular-arc graph implies that $G^{k+2}$ is a circular-arc graph.*

The result of [Lubiw, 1982] and [Dahlhaus & Duchet, 1987] that strongly chordal graphs (section 7.12) are closed under powers is strengthened in [Raychaudhuri, 1992a].

**Theorem 7.57 (Raychaudhuri)** *The intersection class of strongly chordal graphs is strongly closed under powers.*

[Lundgren, Merz, & Rasmussen, 1993] investigates characterizing graphs whose squares are interval graphs. [Brandstädt, Dragan, Chepoi, & Voloshin, 1994] shows that the intersection class of clique graphs of chordal graphs (section 7.5) is closed under powers.

[Jamison, to appear] shows that every power of a block graph is chordal, where a *block graph* is any graph isomorphic to the intersection graph of the vertex sets of all the blocks—maximal 2-connected subgraphs—of a graph. (Block graphs are not to be confused with the block-intersection graphs of designs in section 7.8).) [Harary, 1963] characterizes block graphs as the graphs in which every block is complete.

## 7.11   Sphere-of-Influence Graphs

Suppose $\mathcal{X}$ is any finite set of points in $\mathbb{R}^2$ and each $x \in \mathcal{X}$ is associated with the open ball centered at $x$ with radius equal to the smallest distance from $x$ to any other point of $\mathcal{X}$. A graph is a *sphere-of-influence graph* if it is isomorphic to the intersection graph of such open balls for some $\mathcal{X} \subset \mathbb{R}^2$; [Lipman, 1992] shows that the points $x$ can always be assumed to be lattice points of the plane. *Closed sphere-of-influence graphs* are defined similarly using closed balls. The geometric delicacy involved in these definitions can be glimpsed in Figure 7.18, which shows that the cycle $C_5$ is a sphere-of-influence graph. Note the tangency of the upper left and upper right circles; by Theorem 7.59, $C_5$ is not a closed sphere-of-influence graph.

Sphere-of-influence graphs grew out the [Toussaint, 1988] discussion of pattern recognition and computer vision; [Jaromczyk & Toussaint, 1992] contains references and discussion of general *proximity graphs* of this sort.

Much work has been done attempting to characterize which graphs are (closed) sphere-of-influence graphs, but the general problems remain open. One awkward fact is that the classes of all sphere-of-influence graphs and of all closed sphere-of-influence graphs are not closed under induced subgraphs, and so are not intersection classes in the sense of section 1.2.

A $\{G_1, \ldots, G_k\}$-*factor* of a graph is a spanning subgraph consisting of vertex-disjoint copies of graphs isomorphic to graphs each of which is isomorphic to one of $G_1, \ldots, G_k$; a $\{K_2\}$-factor is a *perfect matching*. The following results are from [Jacobson, Lipman, & McMorris, 1995] and [Michael & Quint, to appear]. (Recall that $P_n$ denotes a path with $n - 1$ edges.)

**Theorem 7.58 (Jacobson, Lipman, & McMorris)** *A tree is a sphere-of-influence graph if and only if it has a perfect matching and is a closed sphere-of-influence graph if and only if it has a $\{P_2, P_3\}$-factor.*

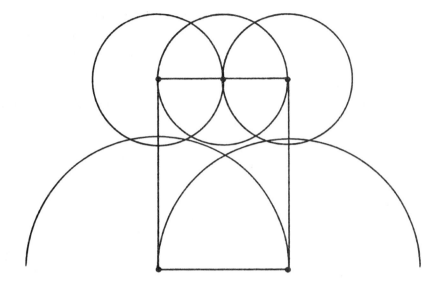

Figure 7.18: *Open balls showing that $C_5$ is a sphere-of-influence graph.*

**Theorem 7.59 (Jacobson, Lipman & McMorris)** *Every triangle-free closed sphere-of-influence graph has a perfect matching.*

**Theorem 7.60 (Michael & Quint)** *Every triangle-free sphere-of-influence graph or triangle-free closed sphere-of-influence graph is planar.*

Even the status of complete graphs is not fully known: there is a gap between $K_8$, which is known to be a sphere-of-influence graph, and $K_{12}$, which is known not to be by the following result of [Kézdy & Kubicki, 1997]. [Harary, Jacobson, Lipman, & McMorris, 1993] conjectures that $K_9$ is not a sphere-of-influence graph.

**Theorem 7.61 (Kézdy & Kubicki)** $K_{12}$ *is not a sphere-of-influence graph.*

[Michael & Quint, 1994] surveys sphere-of-influence graphs and has a bibliography that is complete through 1993. It also shows how the work on sphere-of-influence graphs finds natural expression when extended to arbitrary metric spaces. [Chen, Gould, Jacobson, Schelp, & West, 1992] and [Harary, Jacobson, Lipman, & McMorris, 1994] present a related notion of *influence graph,* replacing euclidean distance with the usual graph-theoretic distance metric. [Guibas, Pach, & Sharir, 1994] investigates sphere-of-influence graphs in higher dimensions. [Holm & Bogart, to appear] studies min- and sum-tolerance sphere-of-influence graphs.

**Theorem 7.62 (Holm & Bogart)** *Every tree is both a min-tolerance sphere-of-influence graph and a sum-tolerance sphere-of-influence graph.*

[Lipman, 1996] studies max-tolerance sphere-of-influence graphs and shows that not every tree is a max-tolerance sphere-of-influence graph.

**Theorem 7.63 (Lipman)** *The complete bipartite graph $K_{1,n}$ is a max-tolerance sphere-of-influence graph if and only if $1 \leq n \leq 5$.*

[McMorris & Wang, to appear] initiates the study of *sphere-of-attraction graphs*, motivated by applications to marketing in [Crama, Hansen, & Jaumard, 1995]. These are defined by two sets $\mathcal{X}_c$ and $\mathcal{X}_p$ of points, whose elements can be thought of as, respectively, "customers" and "products," with balls (open or closed) centered at points of $\mathcal{X}_c$ where each radius is determined by the shortest distance to a point in $\mathcal{X}_p$. McMorris & Wang prove that, in $\mathbb{R}^1$, closed sphere-of-attraction graphs are proper interval graphs, and they give a forbidden subgraph characterization of the $\mathbb{R}^1$ closed sphere-of-attraction graphs. They also show that every proper interval graph is an $\mathbb{R}^2$ closed sphere-of-attraction graph.

**Theorem 7.64 (McMorris & Wang)** *A triangle-free graph is an $\mathbb{R}^2$ sphere-of-attraction graph (or, equivalently, a closed $\mathbb{R}^2$ sphere-of-attraction graph) if and only if it is planar.*

## 7.12   Strongly Chordal Graphs

A chordal graph is *strongly chordal* if it has the additional property that every cycle $C$ of even length at least six has a chord that divides $C$ into two odd-length paths. Strongly chordal graphs form an intermediate family between the families of interval graphs and chordal graphs. They have been particularly important because certain graph-theoretical problems have efficient computational solutions for subfamilies of the family of strongly chordal graphs. However, we will focus on structural properties, primarily from the fundamental paper [Farber, 1983].

Define a *trampoline*, sometimes called a *k-sun*, to be a graph formed from an even-length cycle $v_1, \ldots, v_{2k}, v_1$ by adding edges between even-subscripted vertices so that $\{v_2, v_4, \ldots, v_{2k}\}$ induces a complete subgraph.

Figure 7.19: *A strongly chordal graph and its clique tree.*

Figure 7.20: *A chordal, but not strongly chordal, graph and its clique tree.*

For instance, the graphs on the left in Figures 7.11 and Figure 7.14 are trampolines. (Warning: Some authors use "trampoline" and "$k$-sun" with the same meaning except with "complete" replaced with "chordal"; the two meanings are equivalent in the following characterization by results in [Farber, 1983] and [Chang & Nemhauser, 1984].)

**Theorem 7.65 (Farber)** *A graph is strongly chordal if and only if it is chordal and contains no induced trampoline.*

Define a vertex $v \in V(G)$ to be *simple* if, for every $u, w \in N[v]$, the closed neighborhoods $N[u]$ and $N[w]$ are comparable by inclusion. For instance, vertex 1 in Figure 7.19 is simple since $N[1] = \{1, 2, 4, 5\}$ and $N[1] \subseteq N[2] \subseteq N[4] = N[5]$; the graph in Figure 7.20 has no simple vertex. Call an ordering $\langle v_1, \ldots, v_n \rangle$ of all the vertices of $G$ a *simple elimination ordering* of $G$ if, for each $i \in \{1, \ldots, n\}$, $v_i$ is a simple vertex of the subgraph induced by $v_i, \ldots, v_n$. The vertices in the strongly chordal graph in Figure 7.19 are numbered in a simple elimination ordering. The following theorem parallels the perfect elimination ordering characterization of chordal graphs in Theorem 2.5. ([Farber, 1983] also introduces a different, somewhat less simple notion of "strong elimination ordering" that also characterizes strongly chordal graphs.)

**Theorem 7.66 (Farber)** *A graph is strongly chordal if and only if it has a simple elimination ordering.*

[Chang & Nemhauser, 1984] observes that a graph is chordal if and only if, for every cycle $C$ of length greater than three, there is a triangle consisting of two edges of $C$ and one chord of $C$. [Dahlhaus, Manuel, & Miller, 1998] proves that a chordal graph is strongly chordal if and only if, for every cycle $C$ of length greater than five, there is a triangle consisting of one edge of $C$ and two chords of $C$.

[Brandstädt, Dragan, Chepoi, & Voloshin, 1994] shows that a graph is strongly chordal if and only if every induced subgraph is the clique graph of a chordal graph (section 7.5). [Bandelt & Prisner, 1991] shows that a graph $G$ is strongly chordal if and only if $G = K(H)$ where $H$ is another strongly chordal graph. [Raychaudhuri, 1988] gives an algorithm for the intersection number of strongly chordal graphs.

Strongly chordal graphs are intimately related to totally balanced hypergraphs in [Anstee & Farber, 1984]; see also [Lubiw, 1987]. The second part of the following theorem also appears in [Brouwer, Duchet, & Schrijver, 1983], which contains the very similar characterization of chordal bipartite graphs stated in Theorem 7.22. [Ma & Wu, 1990] shows that Theorem 7.67 is also true when $\mathcal{E}$ is the family of all minimal vertex separators of a chordal graph $G$.

**Theorem 7.67 (Farber)** *A graph $G$ is strongly chordal if and only if the hypergraph $(V(G), \mathcal{E})$, with $\mathcal{E}$ the family of all maxcliques of $G$, is totally balanced; moreover, the same is true when $\mathcal{E}$ is the family of all closed neighborhoods of vertices of $G$.*

Call a clique tree $T$ a *strong clique tree representation* for a graph $G$ when there are *not* vertices $v_1, \ldots, v_k \in V(G)$ and $Q_1, \ldots, Q_k \in V(T)$, $k \geq 3$, such that $Q_1$ is in $T_1$ and $T_2$, $Q_2$ is in $T_2$ and $T_3$, $\ldots$, and $Q_k$ is in $T_k$ and $T_1$, with no $Q_i$ in any other $T_j$ for $i, j \in \{1, \ldots, k\}$. For instance, the clique tree in Figure 7.20 can be seen not to be a strong clique tree by taking $v_1 = 2$, $v_2 = 5$, $v_3 = 7$, and $Q_1 = 1245$, $Q_2 = 4567$, $Q_3 = 2347$ (in other words, $T$ is not a strong clique tree because of the "cyclic" arrangement of the subtrees $T_2$, $T_5$, and $T_7$). This is expressed in terms of either of the hypergraphs $(V(G), V(T))$ and $(V(T), \{T_v : v \in V(G)\})$ not being strongly balanced. The following characterization, paralleling Theorem 2.1, follows from Theorem 7.67, Exercise 2.20, and Corollary 2.10.

**Theorem 7.68** *A graph is strongly chordal if and only if it has a strong clique tree representation.*

[McKee, to appear(b)] characterizes such strong clique tree representations $T$ of $G$ by, for all $v \in V(G)$, each $T_v', T_v'', \ldots$ being connected, where (mimicking section 2.1) $T'$ is defined to be any maximum spanning tree of the weighted intersection graph $K^w(E(T))$ of the set $\{Q : Q \in E(T)\}$, $T''$ is defined similarly in terms of $K^w(E(T'))$, and so on.

Reflecting their connection to totally balanced hypergraphs, strongly chordal graphs are also intimately related to chordal bipartite graphs in [Dahlhaus, 1989], [Hoffman, Kolen, & Sakarovitch, 1985], [Hammer, Maffray, & Preissmann, 1989], and [Brandstädt, 1991]. For instance, Brandstädt's paper includes proofs of the following two results (which should be compared with Exercise 5.11 and Theorem 5.5).

**Theorem 7.69 (Brandstädt and Müller)** *A graph $G$ on $V(G) = \{v_1, \ldots, v_n\}$ is strongly chordal if and only if $B(G)$ is chordal bipartite, where $B(G)$ is defined to be the bipartite graph on $V(B(G)) = \{x_1, \ldots, x_n; y_1, \ldots, y_n\}$ with $x_i y_j \in E(B(G))$ exactly when either $i = j$ or $v_i v_j \in E(G)$.*

**Theorem 7.70 (Dahlhaus)** *A bipartite graph $G$ is chordal bipartite if and only if the split graph obtained from $G$ by making one of its two color classes complete is strongly chordal.*

Paralleling Theorem 7.68, [McKee, to appear(b)] also defines a "strong neighborhood tree representation" $T$ in the same way as strong clique tree representations, as a maximum spanning tree $T$ of the weighted intersection graph $K^w(\mathcal{N}(G))$ of the set $\mathcal{N}(G) = \{N(v) : v \in V(G)\}$ of all open neighborhoods of vertices of $G$ such that every $T_v, T_v', T_v'', \ldots$ is connected. Then, related to Theorem 7.22 or 7.70, a graph is chordal bipartite if and only if it has a "strong neighborhood tree representation."

# Bibliography

**B. D. Acharya & M. Las Vergnas (1982).**
Hypergraphs with cyclomatic number zero, triangulated graphs, and an inequality. *J. Combin. Theory Ser. B* **33**, 52–56. (Cited in §2.1.)

**B. D. Acharya & M. N. Vartak (1973).**
Open Neighborhood Graphs. *Research Report of the Indian Institute of Technology*, Bombay. (Cited in §4.1.2.)

**M. Ackerman, F. R. McMorris, & S. Seif (1993).**
Chordal intersection graphs of semigroups. *Congr. Numer.* **93**, 45–49. (Cited in §7.8.)

**R. Agarwala & D. Fernández-Baca (to appear).**
Fast and simple algorithms for perfect phylogeny and triangulating colored graphs. (Cited in §2.4.1.)

**B. Alspach & D. Hare (1991).**
Edge-pancyclic block-intersection graphs. *Discrete Math.* **97**, 17–24. (Cited in §7.8.)

**R. Alter & C. C. Wang (1977).**
Uniquely intersectable graphs. *Discrete Math.* **18**, 217–226. (Cited in §1.3.)

**C. A. Anderson (1995).**
Loop and cyclic niche graphs. *Linear Algebra Appl.* **217**, 5–13. (Cited in §4.2.)

**C. A. Anderson, K. F. Jones, J. R. Lundgren, & T. A. McKee (1990).**
Competition multigraphs and the multicompetition number. *Ars Combin.* **29B**, 185–192. (Cited in §6.2.)

C. A. Anderson, K. F. Jones, J. R. Lundgren, & S. Seager (1991).
A suggestion for new niche numbers for graphs. *Congr. Numer.* **81**, 23–32. (Cited in §4.2.)

C. A. Anderson, L. Langley, J. R. Lundgren, P. A. McKenna, & S. K. Merz (1994). New classes of $p$-competition graphs and $\phi$-tolerance competition graphs. *Congr. Numer.* **100**, 97–107. (Cited in §6.1, 6.3.)

T. Andreae, U. Hennig, & A. Parra (1993).
On a problem concerning tolerance graphs. *Discrete Appl. Math.* **46**, 73–78. (Cited in §6.3.)

R. P. Anstee & M. Farber (1984).
Characterization of totally balanced matrices. *J. Algorithms* **5**, 215–230. (Cited in §7.12.)

J. E. Atkins & M. Middendorf (1996).
On physical mapping and the consecutive ones property for sparse matrices. *Discrete Appl. Math.* **71**, 23–40. (Cited in §3.4.1.)

M. Bakonyi (1992).
On Gaussian elimination and determinant formulas for matrices with chordal inverses. *Bull. Austral. Math. Soc.* **46**, 435–440. [Corrigendum, **49**, p. 175.] (Cited in §2.4.3.)

M. Bakonyi & A. Bono (1997).
Several results on chordal bipartite graphs. *Czechoslovak Math. J.* **47**, 577–583. (Cited in §7.3, 7.8.)

M. Bakonyi & C. R. Johnson (1995).
The Euclidean distance matrix completion problem. *SIAM J. Matrix Anal. Appl.* **16**, 646–654. (Cited in §2.4.3.)

M. Bakonyi & C. R. Johnson (1996).
Algebraic characterizations of chordality. *Linear and Multilinear Algebra* **40**, 187–191. (Cited in §2.2, 2.4.3.)

R. Balakrishnan & P. Paulraja (1983).
Powers of chordal graphs. *J. Austral. Math. Soc. Ser. A* **35**, 211–217. (Cited in §7.10.)

**H.-J. Bandelt & E. Prisner (1991).**
Clique graphs and Helly graphs. *J. Combin. Theory Ser. B* **51**, 34–45. (Cited in §2.1, 7.5, 7.12.)

**J. Bang-Jensen & P. Hell (1994).**
On chordal proper circular arc graphs. *Discrete Math.* **128**, 395–398. (Cited in §7.1.)

**W. W. Barrett & C. R. Johnson (1984).**
Determinantal formulae for matrices with sparse inverses. *Linear Algebra Appl.* **56**, 73–88. (Cited in §2.4.3.)

**W. W. Barrett, C. R. Johnson, & M. Lundquist (1989).**
Determinantal formulae for matrix completions associated with chordal graphs. *Linear Algebra Appl.* **121**, 265–289. (Cited in §2.1, 2.4.3.)

**A. Batbedat (1990).**
*Les approches pyramidales dans la classification arboree.* Masson, Paris. (Cited in §7.5.)

**C. Beeri, R. Fagin, D. Maier, & M. Yannakakis (1983).**
On the desirability of acyclic database schemes. *J. Assoc. Comput. Mach.* **30**, 479–513. (Cited in §2.4.2.)

**L. W. Beineke (1968).**
Derived graphs and digraphs. In *Beiträge zur Graphentheorie* (H. Sachs, H.-J. Voss, & H. Walther, eds.), Teubner, Leipzig; pp. 17–33. (Cited in §1.5.)

**L. W. Beineke & C. M. Zamfirescu (1982).**
Connection digraphs and second order line graphs. *Discrete Math.* **39**, 237–254. (Cited in §7.2.)

**S. Bellantoni, I. B. A. Hartman, T. Przytycka, & S. Whitesides (1993).** Grid intersection graphs and boxicity. *Discrete Math.* **114**, 41–49. (Cited in §7.1.)

**C. Benzaken, Y. Crama, P. Duchet, P. L. Hammer, & F. Maffray (1990).** More characterizations of triangulated graphs. *J. Graph Theory* **14**, 413–422. (Cited in §2.2.)

**S. Benzer (1959).**
On the topology of the genetic fine structure. *Proc. Nat. Acad. Sci. U.S.A.* **45**, 1607–1620. (Cited in §3.4.1.)

**C. Berge (1989).**
*Hypergraphs. Combinatorics of Finite Sets.* North–Holland, Amsterdam. (Cited in Preface and §1.6)

**D. J. Bergstrand & K. F. Jones (1988).**
On upper bound graphs of partially ordered sets. *Congr. Numer.* **66**, 185–193. (Cited in §4.4.)

**D. J. Bergstrand & K. F. Jones (1989).**
Graphs that are both the upper and lower bound graphs of a poset. *Ars Combin.* **28**, 109–121. (Cited in §4.4.)

**D. J. Bergstrand, K. F. Jones, & W. R. Sherman (to appear).**
Posets with isomorphic upper and lower bound graphs. (Cited in §4.4.)

**P. A. Bernstein & N. Goodman (1981).**
Power of natural semijoins. *SIAM J. Comput.* **10**, 751–771. (Cited in §2.1.)

**J. R. S. Blair & B. Peyton (1993).**
An introduction to chordal graphs and clique trees. In *Graph Theory and Sparse Matrix Computation* (A. George, J. R. Gilbert, & J. W. H. Liu, eds.), Springer, New York; pp. 1–29. (Cited in beginning of §2, 2.2.)

**J. R. S. Blair & B. Peyton (1994).**
On finding minimum-diameter clique trees. *Nordic J. Comput.* **1**, 173–201. (Cited in §2.1.)

**H. Bodlaender, M. Fellows, & T. Warnow (1992).**
Two strikes against perfect phylogeny. In *Proceedings of the 19th International Colloquium on Automata, Languages, and Programming* (W. Kuich, ed.), [*Lecture Notes in Computer Science* **623**], Springer, New York; pp. 273–283. (Cited in §2.4.1.)

**H. L. Bodlaender & B. de Fluiter (1996).**
On intervalizing $k$-colored graphs for DNA physical mapping. *Discrete Appl. Math.* **71**, 55–77. (Cited in §3.4.1.)

**H. Bodlaender & T. Kloks (1993).**
A simple linear time algorithm for triangulating three-colored graphs. *J. Algorithms* **15**, 160–172. (Cited in §2.4.1.)

**K. P. Bogart, P. C. Fishburn, G. Isaak, & L. J. Langley (1995).**
Proper and unit tolerance graphs. *Discrete Appl. Math.* **60**, 99–117. (Cited in §6.3.)

**K. S. Booth & G. S. Leuker (1976).**
Testing for the consecutive ones property, interval graphs, and graph planarity using PQ-trees. *J. Comput. Systems. Sci.* **13**, 335–379. (Cited in §3.1.)

**C. F. Bornstein & J. L. Szwarcfiter (1995).**
On clique convergent graphs. *Graphs Combin.* **11**, 213–220. (Cited in §7.5.)

**J. Bosák (1964).**
The graphs of semigroups. In *Theory of Graphs and Its Applications* (M. Fiedler, ed.) Academic Press, New York; pp. 119–125. (Cited in §7.8.)

**J. Bosák (1975).**
Graphs of algebras and algebraic graphs. In *Recent Advances in Graph Theory* (M. Fiedler, ed.) Academia, Prague; pp. 93–98. (Cited in §7.8.)

**A. Bouchet (1994).**
Circle graph obstructions. *Discrete Math.* **60**, 107–144. (Cited in §7.4.)

**A. Brandstädt (1991).**
Classes of bipartite graphs related to chordal graphs. *Discrete Appl. Math.* **32**, 51–60. (Cited in §7.2, 7.12.)

**A. Brandstädt (1993).**
*Special graph classes—A survey.* Schriften reihe des Fachbereichs Mathematikik **SM-DU-1993**, Universität Duisburg, 1993. (Cited in Preface, beginning of §7, §7.3.)

**A. Brandstädt, V. D. Chepoi, & F. F. Dragan (1995).**
The algorithmic use of hypertree structure and maximum neighbourhood orderings. In *Graph-Theoretic Concepts in Graph Theory* (E. W. Mayr, G. Schmidt, & G. Tinhofer, eds.) [*Lecture Notes in Computer Science* **903**] Springer, Berlin; pp. 65–80. (Cited in §7.5.)

A. Brandstädt, V. D. Chepoi, & F. F. Dragan (1996).
Perfect elimination orderings of chordal powers of graphs. *Discrete Math.* **158**, 273–278. (Cited in §7.10.)

A. Brandstädt, F. F. Dragan, V. D. Chepoi, & V. I. Voloshin (1994). Dually chordal graphs. In *Graph-Theoretic Concepts in Graph Theory* (J. van Leeuwen, ed.) [*Lecture Notes in Computer Science* **790**] Springer, Berlin; pp. 237–251. (Cited in §7.5, 7.10, 7.12.)

A. Brandstädt, V. B. Le & J. Spinrad (to appear).
*Graph classes — A Survey.* Society for Industrial and Applied Mathematics, Philadelphia. (Cited in Preface, beginning of §7)

E. O. Brauner, R. A. Brualdi, & E. S. N. Sneyd (1995).
Pseudo-interval graphs. *J. Graph Theory* **20**, 309–318. (Cited in §7.2.)

R. C. Brigham, R. D. Dutton, & F. R. McMorris (1992).
On the relationship between $p$-edge and $p$-vertex clique covers. *Vishwa Internat. J. Graph Theory* **1**, 133–140. (Cited in §6.1.)

R. C. Brigham, R. D. Dutton, & F. R. McMorris (1993).
On $p$-edge clique covers of graphs. *Congr. Numer.* **93**, 149–157. (Cited in §6.1.)

R. C. Brigham, F. R. McMorris, & R. P. Vitray (1995).
Tolerance competition graphs. *Linear Algebra Appl.* **217**, 41–52. (Cited in §6.3.)

R. C. Brigham, F. R. McMorris, & R. P. Vitray (1996).
Two-$\phi$-tolerance competition graphs. *Discrete Appl. Math.* **66**, 101–108. (Cited in §6.3.)

R. C. Brigham, F. R. McMorris, & R. P. Vitray (to appear).
Bipartite graphs and absolute difference tolerances. *Ars Combin.*, (Cited in §6.3.)

A. E. Brouwer, P. Duchet, & A. Schrijver (1983).
Graphs whose neighborhoods have no special cycles. *Discrete Math.* **47**, 177–182. (Cited in §7.3, 7.12.)

P. Buneman (1974).
A characterisation of rigid circuit graphs. *Discrete Math.* **9**, 205–212. (Cited in §2.1, 2.4.1.)

**S. Bylka & J. Komar (1997).**
Intersection properties of line graphs. *Discrete Math.* **164**, 33–45.
(Cited in §6.2.)

**L. Cai, D. Corneil, & A. Proskurowski (1996).**
A generalization of line graphs: $(X, Y)$-intersection graphs. *J. Graph Theory* **21**, 267–287. (Cited in §1.5.)

**S. Chaiken, N. V. Murray, & E. Rosenthal (1989).**
An application of $P_4$-free graphs in theorem proving. In *Combinatorial Mathematics: Proceedings of the Third International Conference* (G. S. Bloom, R. Graham, & J. Malkevitch, eds.) [*Ann. New York Acad. Sci.* **555**] N. Y. Acad. Sci., New York; pp. 106–121. (Cited in §7.9.)

**R. Chandrasekaran & A. Tamir (1982).**
Polynomially bounded algorithms for locating $p$-centers on a tree. *Math. Programming* **22**, 304–315. (Cited in §2.4.)

**G. J. Chang & G. L. Nemhauser (1984).**
The $k$-domination and $k$-stability problems on sun-free chordal graphs. *SIAM J. Alg. Discrete Methods* **3**, 332–345. (Cited in §7.12.)

**Y. W. Chang, M. S. Jacobson, C. L. Monma, & D. B. West (1993).**
Subtree and substar intersection numbers. *Discrete Appl. Math.* **44**, 205–220. (Cited in §7.1.)

**G. Chartrand & L. Lesniak (1996).**
*Graphs & Digraphs*, Third Edition. Chapman & Hall, London. (Cited in Preface, §1.1.)

**F. Cheah & D. G. Corneil (1996).**
On the structure of trapezoid graphs. *Discrete Appl. Math.* **66**, 109–133. (Cited in §7.1.)

**G. Chen, R. J. Gould, M. S. Jacobson, R. H. Schelp, & D. B. West (1992).** A characterization of influence graphs of a prescribed girth. *Vishwa Internat. J. Graph Theory* **1**, 77–81. (Cited in §7.11.)

**B.-L. Chen & K.-W. Lih (1990).**
Diameters of iterated clique graphs of chordal graphs. *J. Graph Theory* **14**, 391–396. (Cited in §2.1.)

G. A. Cheston, E. O. Hare, S. T. Hedetniemi, & R. C. Laskar
(1988). Simplicial graphs. *Congr. Numer.* **67**, 105–113. (Cited in
§4.4.)

R. Christensen (1990).
*Log-Linear Models.* Springer, New York. (Cited in §2.4.4.)

F. R. K. Chung & D. Mumford (1994).
Chordal completions of planar graphs. *J. Combin. Theory Ser. B* **62**,
96–106. (Cited in §2.4.2.)

M. S. Chung & D. B. West (1994).
The $p$-intersection number of a complete bipartite graph and orthogonal double coverings of a clique. *Combinatorica* **14**, 453–461. (Cited
in §6.1.)

V. Chvátal & P. L. Hammer (1973).
Set-Packing and Threshold Graphs. Res. Report **CORR 73-21**, University of Waterloo. (Cited in beginning of §5, §5.1, 5.4, 7.9.)

V. Chvátal & P. L. Hammer (1977).
Aggregations of inequalities. In *Studies in Integer Programming* (P.
L. Hammer, E. L. Johnson, B. H. Korte, & G. L. Nemhauser, eds.),
[*Ann. Discrete Math.* **1**], North–Holland, Amsterdam, pp. 145–162.
(Cited in §5.4.)

B. N. Clark, C. J. Colbourn, & D. S. Johnson (1990).
Unit disk graphs. *Discrete Math.* **86**, 165–177. (Cited in §7.1.)

O. Cogis (1984).
Ferrers digraphs and threshold graphs. *Discrete Math.* **38**, 33–46.
(Cited in §5.3.)

J. E. Cohen (1978).
*Food Webs and Niche Space.* Princeton University Press, Princeton,
NJ. (Cited in §4.2, 4.3, 6.2.)

J. E. Cohen, F. Briand, & C. M. Newman (1990).
*Community Food Webs: Data and Theory.* Springer, Heidelberg.
(Cited in §4.3.)

J. E. Cohen & Z. J. Palka (1990).
A stochastic theory of community food webs. *V.* Intervality and

triangulation in the trophic-niche overlap graph. *Amer. Naturalist* **135**, 435–463. (Cited in §4.3.)

**C. H. Coombs & J. E. K. Smith (1973).**
On the detection of structures in attitudes and developmental processes. *Psych. Rev.* **80**, 337–351. (Cited in §3.4.2.)

**D. G. Corneil & P. A. Kamula (1987).**
Extensions of permutation and interval graphs. *Congr. Numer.* **58**, 267–275. (Cited in §7.1, 7.6.)

**D. G. Corneil, H. Kim, S. Natarajan, S. Olariu, & A. Sprague (1995).** Simple linear time recognition of unit interval graphs. *Inform. Process. Lett.* **55**, 99–104. (Cited in §3.3.)

**D. G. Corneil, H. Lerchs, & L. Stewart Burlingham (1981).**
Complement reducible graphs. *Discrete Appl. Math.* **3**, 163–174. (Cited in §7.9.)

**D. G. Corneil, S. Olariu, & L. Stewart (1997).**
Asteroidal triple-free graphs. *SIAM J. Discrete Math.* **10**, 399–430. (Cited in §7.6.)

**D. G. Corneil, S. Olariu, & L. Stewart (1998).**
The ultimate interval graph recognition algorithm? Extended abstract in *Proceedings of the Ninth Annual ACM-SIAM Symposium on Discrete Algorithms (SODA)* (1998), pp. 175–180. (Cited in §3.1.)

**D. G. Corneil, Y. Perl, & L. K. Stewart (1985).**
A linear recognition algorithm for cographs. *SIAM J. Comput.* **14**, 926–934. (Cited in §7.9.)

**M. B. Cozzens & R. Leibowitz (1984).**
Threshold dimensions of graphs. *SIAM J. Alg. Discrete Methods* **5**, 579–595. (Cited in §5.4.)

**M. B. Cozzens & R. Leibowitz (1987).**
Multidimensional scaling and threshold graphs, *J. Math. Psych.* **31**, 179–191. (Cited in §5.4.)

**M. B. Cozzens & N. V. R. Mahadev (1989).**
Consecutive one's properties for matrices and graphs including variable diagonal entries. In *Applications of Combinatorics and Graph Theory*

*to the Biological and Social Sciences* (F. S. Roberts, ed.), Springer, New York; pp. 75–94. (Cited in §7.1.)

**M. B. Cozzens & F. S. Roberts (1989).**
On dimensional properties of graphs. *Graphs Combin.* **5**, 29–46. (Cited in §7.1.)

**W. L. Craine (1994).**
Characterizations of fuzzy intersection graphs. *Fuzzy Sets and Systems* **68**, 181–193. (Cited in §7.8.)

**Y. Crama, P. Hansen, & B. Jaumard (1995).**
Complexity of product positioning and ball intersection problems. *Math. Oper. Res.* **20**, 889–891. (Cited in §7.11.)

**B. Csákány & G. Pollák (1969).**
The graph of subgroups of a finite group. [Russian; see Math. Rev. **40** #2573] *Czechoslovak Math. J.* **19(94)**, 242–247. (Cited in §7.8.)

**E. Dahlhaus (1989).**
*Chordale Graphen im Desonderen Hinblick auf parallelle Algorithmen.* Habilitationsschrift, Universität Bonn. (Cited in §7.12.)

**E. Dahlhaus & P. Duchet (1987).**
On strongly chordal graphs. *Ars Combin.* **24B**, 23–30. (Cited in §7.10.)

**E. Dahlhaus, P. D. Manuel, & M. Miller (1998).**
A characterization of strongly chordal graphs. *Discrete Math.* **187**, 269–271. (Cited in §7.12.)

**J. N. Darroch, S. L. Lauritzen, & T. P. Speed (1980).**
Markov fields and log-linear interaction models for contingency tables. *Ann. Statist.* **8**, 522–539. (Cited in §2.4.4.)

**I. Degan, M. C. Golumbic, & R. Y. Pinter (1988).**
Trapezoid graphs and their coloring. *Discrete Appl. Math.* **21**, 35–46. (Cited in §7.1.)

**X. Deng, P. Hell, & J. Huang (1996).**
Linear time representation algorithms for proper circular arc graphs and proper interval graphs. *SIAM J. Comput.* **25**, 390–403. (Cited in §3.3, 7.1.)

**J. S. Deogun & K. Gopalakrishnan (to appear).**
Consecutive retrieval property—Revisited. (Cited in §3.3, 3.4.3)

**D. W. DeTemple, M. J. Dineen, J. M. Robertson, & K. McAvaney (1993).** Recent examples in the theory of partition graphs. *Discrete Math.* **113**, 255–258. (Cited in §7.8.)

**D. W. DeTemple, J. Robertson, & F. Harary (1984).**
Existential partition graphs. *J. Combin. Inform. System Sci.* **9**, 193–196. (Cited in §7.8.)

**R. Diestel (1988).**
Tree-decompositions, tree-representability and chordal graphs. *Discrete Math.* **71**, 181–184. (Cited in §7.7.)

**R. Diestel (1990).**
*Graph Decompositions. A Study in Infinite Graph Theory.* Clarendon Press, Oxford. (Cited in §7.7.)

**G. A. Dirac (1961).**
On rigid circuit graphs. *Abh. Math. Sem. Univ. Hamburg* **25**, 71–76. (Cited in §2.2.)

**F. Dragan (1993).**
HT-graphs: Centers, connected $r$-domination and Steinertrees. *Comput. Sci. J. Moldova* **1**, 64–83. (Cited in §7.5.)

**P. Duchet (1978).**
Propriété de Helly et problèmes de représentation. *Colloques Internat. CNRS* **260**, 117–118. (Cited in §2.3, 3.2, 7.1.)

**P. Duchet (1984).**
Classical perfect graphs. An introduction with emphasis on triangulated and interval graphs. In *Topics in Perfect Graphs* (C. Berge & V. Chvátal, eds.), [*North-Holland Math. Stud.* **88**], North-Holland, Amsterdam. *Ann. Discrete Math.* **21**, 67–96. (Cited in §3.2, 6.2, 7.1, 7.10.)

**P. Duchet (1995).**
Hypergraphs. In *Handbook of Combinatorics* (R. L. Graham, M. Grotschel, & L. Lovász,, eds.), Elsevier, Amsterdam; Vol. 1, pp. 381–432. (Cited in §1.6, 3.2.)

**B. Dushnik & E. W. Miller (1941).**
Partially ordered sets. *Amer. J. Math.* **63**, 600–610. (Cited in §7.6.)

**R. D. Dutton & R. C. Brigham (1983).**
A characterization of competition graphs. *Discrete Appl. Math.* **6**, 315–317. (Cited in §4.2.)

**N. Eaton (1997).**
Intersection representations of complete unbalanced bipartite graphs. *J. Combin. Theory Ser. B* **71**, 123–129. (Cited in §6.1.)

**N. Eaton, R. J. Gould, & V. Rödl (1996).**
The representation of a graph by set intersections. *J. Graph Theory* **21**, 377–392. (Cited in §6.1.)

**N. Eaton & D. A. Grable (1996).**
Set intersection representations for almost all graphs. *J. Graph Theory* **23**, 309–320. (Cited in §6.1.)

**G. Ehrlich, S. Even, & R. E. Tarjan (1976).**
Intersection graphs of curves in the plane. *J. Combin. Theory Ser. B* **21**, 8–20. (Cited in §7.1.)

**E. S. Elmallah & L. K. Stewart (1993).**
Independence and domination in polygon graphs. *Discrete Math.* **44**, 65–77. (Cited in §7.4.)

**H. Era & M. Tsuchiya (1991).**
On intersection numbers of graphs. In *Graph Theory, Combinatorics, Algorithms and Applications* (Y. Alavi, F. R. K. Chung, R. L. Graham, & D. F. Hsu, eds.) Society for Industrial and Applied Mathematics, Philadelphia; pp. 545–556. (Cited in §1.3.)

**H. Era & M. Tsuchiya (1997).**
Remarks on relations between upper bound graphs and double bound graphs. *Proc. School Sci. Tokai Univ.* **32**, 1–6. (Cited in §4.4.)

**H. Era & M. Tsuchiya (1998).**
On upper bound graphs whose complements are also upper bound graphs. *Discrete Math.* **179**, 103–109. (Cited in §4.4.)

**P. Erdős, C. D. Godsil, S. G. Krantz, & T. Parsons (1988).**
Intersection graphs for families of balls in $\mathbb{R}^n$. *Europ. J. Math.* **9**, 501–505. (Cited in §7.1.)

**P. Erdős, A. W. Goodman, & L. Pósa (1966)**.
The representation of a graph by set intersections. *Canad. J. Math.*
**18**, 106–112. (Cited in §1.3.)

**F. Escalante (1973)**.
Über iterierte Clique-Graphen. *Abh. Math. Sem. Univ. Hamburg* **39**,
59–68. (Cited in §7.5.)

**F. Escalante, L. Montejano, & T. Rojano (1974)**.
Characterization of $n$-path graphs and of graphs having $n$th root.
*J. Combin. Theory Ser. B* **16**, 282–289. (Cited in §4.1.2.)

**E. M. Eschen, R. B. Hayward, J. P. Spinrad, & R. Sritharan (to
appear)**. Weakly triangulated comparability graphs. (Cited in §7.3,
7.6.)

**E. M. Eschen & J. P. Spinrad (1993)**.
An $O(n^2)$ algorithm for circular-arc graph recognition. In *Proceedings
of the Fourth Annual ACM-SIAM Symposium on Discrete Algorithms*.
Association for Computing Machinery, New York; pp. 128–137. (Cited
in §7.1, 7.3.)

**K. P. Eswaran (1975)**.
Faithful representation of a family of sets by a set of intervals. *SIAM
J. Comput.* **4**, 56–68. (Cited in §3.4.3.)

**S. Even & A. Itai (1971)**.
Queues, stacks and graphs. In *Theory of Machines and Computations*
(Z. Kohavi & A. Paz, eds.) Academic Press, New York; pp. 71–86.
(Cited in §7.4.)

**M. Farber (1983)**.
Characterizations of strongly chordal graphs. *Discrete Math.* **43**, 173–
189. (Cited in §7.12.)

**M. R. Fellows, M. T. Hallett, & H. T. Wareham (1993)**.
DNA physical mapping: Three ways difficult. In *Algorithms—ESA '93*
(T. Lengauer, ed.), [*Lecture Notes in Computer Science* **726**], Springer,
Berlin; pp. 157–168. (Cited in §3.4.1.)

**S. Felsner (1993)**.
Tolerance graphs and orders. In *Graph-Theoretic Concepts in Computer Science* (E. W. Mayr, ed.), [*Lecture Notes in Computer Science*
**657**], Springer, Berlin; pp. 17–26. (Cited in §6.3, 7.1, 7.6.)

**S. Felsner, R. Müller, & L. Wernisch (1997).**
Trapezoid graphs and generalizations, geometry and algorithms. *Discrete Appl. Math.* **74**, 13–32. (Cited in §7.1.)

**C. M. Fiduccia, E. R. Scheinerman, A. Trenk, & J. S. Zito (1998).**
Dot product representations of graphs. *Discrete Math.* **181**, 113–138. (Cited in §7.8.)

**C. M. H. de Figueiredo, J. Meidanis, & C. P. de Mello (1995).**
A linear-time algorithm for proper interval graph recognition. *Inform. Process. Lett.* **56**, 179–184. (Cited in §3.3.)

**P. C. Fishburn (1970a).**
An interval graph is not a comparability graph. *J. Combin. Theory* **8**, 442–443. (Cited in §3.4.2.)

**P. C. Fishburn (1970b).**
Intransitive indifference with unequal indifference intervals. *J. Math. Psych.* **7**, 144–149. (Cited in §3.4.2.)

**P. C. Fishburn (1985).**
*Interval Orders and Interval Graphs.* Wiley, New York. (Cited in Preface and §3.4.2.)

**W. Fitch (1977).**
On the problem of discovering the most parsimonious trees. *Amer. Nat.* **111**, 223–257. (Cited in §2.4.1.)

**C. Flament (1978).**
Hypergraphes arborés. *Discrete Math.* **21**, 223–226. (Cited in §2.3.)

**C. Flotow (1995).**
On powers of $m$-trapezoid graphs. *Discrete Appl. Math.* **63**, 187–192. (Cited in §7.1.)

**C. Flotow (1996).**
On powers of circular arc graphs and proper circular arc graphs. *Discrete Appl. Math.* **69**, 199–207. (Cited in §7.1, 7.10.)

**C. Flotow (1997).**
Graphs whose powers are chordal and graphs whose powers are interval. *J. Graph Theory* **24**, 323–330. (Cited in §7.10.)

**S. Földes & P. L. Hammer (1977a).**
Split graphs having Dilworth number two. *Canad. J. Math* **29**, 666–672. (Cited in §2.5.)

**S. Földes & P. L. Hammer (1977b).**
Split graphs. *Congr. Numer.* **19**, 311–315. (Cited in §2.5.)

**J. C. Fournier (1978).**
Une caractérisation des graphes de cordes. *C. R. Acad. Sci. Paris* **286A**, 811–813. (Cited in §7.4.)

**K. F. Fraughnaugh, J. R. Lundgren, S. K. Merz, J. S. Maybee, & N. J. Pullman (1995).** Competition graphs of strongly connected and hamiltonian digraphs. *SIAM J. Discrete Math.* **8**, 179–185. (Cited in §4.2.)

**H. de Fraysseix (1984).**
A characterization of circle graphs. *Europ. J. Combin.* **5**, 223–238. (Cited in §7.4.)

**H. B. Frost, M. S. Jacobson, J. A. Kabell, & F. R. McMorris (1990).** Bipartite analogues of split graphs and related topics. *Ars Combin.* **29**, 283–288. (Cited in §7.2.)

**D. R. Fulkerson & O. A. Gross (1965).**
Incidence matrices and interval graphs. *Pacific J. Math.* **15**, 835–855. (Cited in §2.2, 3.1, 6.2.)

**Z. Füredi (1997).**
The $p$-intersection number of the complete bipartite graph. In *The Mathematics of Paul Erdős, Algorithms Combin.* **14**, 86–92. (Cited in §6.1.)

**P. Galinier, M. Habib, & C. Paul (1995).**
Chordal graphs and their clique graphs. In *Graph-Theoretic Concepts in Computer Science* (M. Nagl, ed.), [*Lecture Notes in Computer Science* **1017**], Springer, Berlin, pp. 358–371. (Cited in §2.2.)

**B. Ganter, H.-D. O. F. Gronau, & R. C. Mullin (1994).**
On orthogonal double covers of $K_n$. *Ars Combin.* **37**, 209–221. (Cited in §6.1.)

**E. Gasse (1997).**
A proof of a circle graph characterization. *Discrete Math.* **173**, 277–283. (Cited in §7.4.)

**F. Gavril (1974a).**
The intersection graphs of subtrees in trees are exactly the chordal graphs. *J. Combin. Theory Ser. B* **16**, 47–56. (Cited in §2.1.)

**F. Gavril (1974b).**
Algorithms on circular-arc graphs. *Networks* **4**, 357–369. (Cited in §7.1.)

**F. Gavril (1978).**
A recognition algorithm for the intersection graphs of paths in trees. *Discrete Math.* **23**, 211–227. (Cited in §7.1.)

**F. Gavril (1987).**
Generating the maximum spanning trees of a weighted graph. *J. Algorithms* **8**, 592–597. (Cited in §2.1.)

**F. Gavril (1994).**
Intersection graphs of proper subtrees of unicyclic graphs. *J. Graph Theory* **18**, 615–627. (Cited in §7.1.)

**F. Gavril (1996).**
Intersection graphs of Helly families of subtrees. *Discrete Appl. Math.* **66**, 45–56. (Cited in §7.1.)

**F. Gavril & J. Urrutia (1994).**
Intersection graphs of concatenable subtrees of graphs. *Discrete Appl. Math.* **52**, 195–209. (Cited in §7.1.)

**A. George, J. R. Gilbert, & J. W. H. Liu, Eds. (1993).**
*Graph Theory and Sparse Matrix Computation.* Springer, New York. (Cited in §2.4.3.)

**S. P. Ghosh (1972).**
File organization: The consecutive retrieval property. *Comm. Assoc. Comput. Mach.* **15**, 802–808. (Cited in §3.4.3.)

**S. P. Ghosh (1977).**
*Data Base Organization for Data Management.* Academic Press, New York. (Cited in §3.4.3.)

**S. P. Ghosh (1986).**
*Data Base Organization for Data Management*, Second Edition. Academic Press, Orlando, FL. (Cited in §3.4.3.)

**S. P. Ghosh, Y. Kambayashi, & W. Lipski (1983).**
*Data Base File Organization. Theory and Applications of the Consecutive Retrieval Property.* Academic Press, New York. (Cited in §3.4.3.)

**P. C. Gilmore & A. J. Hoffman (1964).**
A characterization of comparability graphs and of interval graphs. *Canad. J. Math.* **16**, 539–548. (Cited in §3.1.)

**P. W. Goldberg, M. C. Golumbic, H. Kaplan, & R. Shamir (1995).**
Four strikes against physical mapping of DNA. *J. Comput. Bio.* **2**, 139–152. (Cited in §3.4.1.)

**M. C. Golumbic (1978a).**
Trivially perfect graphs. *Discrete Math.* **24**, 105–107. (Cited in §7.9.)

**M. C. Golumbic (1978b).**
A generalization of Dirac's Theorem on triangulated graphs [Second International Conference on Combinatorial Mathematics, New York, 1978.] *Ann. New York Acad. Sci.* **319** (1979) 242–246. (Cited in §7.3.)

**M. C. Golumbic (1978c).**
Threshold graphs and synchronizing parallel processes. In *Combinatorics* (A. Hajnal & V. T. Sós, eds.), Vol. 1, North–Holland, Amsterdam. *Colloq. Math. Soc. János Bolyai* **18**, 419–428. (Cited in §5.4.)

**M. C. Golumbic (1980).**
*Algorithmic Graph Theory and Perfect Graphs.* Academic Press, San Diego. (Cited throughout.)

**M. C. Golumbic (1984).**
Algorithmic aspects of perfect graphs. In *Topics in Perfect Graphs* (C. Berge & V. Chvàtal, eds.), [*North-Holland Math. Stud.* **88**], North–Holland, Amsterdam. *Ann. Discrete Math.* **21**, 301–323. Cited in Preface, §2.2, 2.4.3, 3.4.3.)

**M. C. Golumbic (1988).**
Algorithmic aspects of intersection graphs and representation hypergraphs. *Graphs Combin.* **4**, 307–321. (Cited in §2.4.2.)

**M. C. Golumbic & C. F. Goss (1978).**
Perfect elimination and chordal bipartite graphs, *J. Graph Theory* **2**, 155–163. (Cited in §7.2, 7.3.)

**M. C. Golumbic & R. E. Jamison (1985a).**
The edge intersection graphs of paths in a tree. *J. Combin. Theory Ser. B* **38**, 8–22. (Cited in §6.3, 7.1.)

**M. C. Golumbic & R. E. Jamison (1985b).**
Edge and vertex intersection of paths in a tree. *Discrete Math.* **55**, 151–159. (Cited in §6.3, 7.1.)

**M. C. Golumbic & C. L. Monma (1982).**
A generalization of interval graphs with tolerances. *Congr. Numer.* **35**, 321–331. (Cited in §6.3.)

**M. C. Golumbic, C. L. Monma, & W. T. Trotter (1984).**
Tolerance graphs. *Discrete Appl. Math.* **9**, 157–170. (Cited in §6.3.)

**M. C. Golumbic, D. Rotem, & J. Urrutia (1983).**
Comparability graphs and intersection graphs. *Discrete Math.* **43**, 37–46. (Cited in §7.1, 7.6.)

**M. C. Golumbic & E. R. Scheinerman (1989).**
Containment graphs, posets, and related classes of graphs. In *Combinatorial Mathematics* (G. S. Bloom et al., eds.), *Ann. N. Y. Acad. Sci.* **555**, 192–204. (Cited in §7.6.)

**R. L. Graham & P. Hell (1985).**
On the history of the minimum spanning tree problem. *Ann. Hist. Comput.* **7**, 43–57. (Cited in §2.2.)

**M. J. Greenberg (1980).**
*Euclidean and Non-Euclidean Geometries.* Freeman, San Francisco. (Cited in §7.4.)

**J. R. Griggs (1979).**
Extremal values of the interval number of a graph, II. *Discrete Math.* **28**, 37–47. (Cited in §7.1.)

**J. R. Griggs & D. B. West (1979).**
Extremal values of the interval number of a graph, I. *SIAM J. Alg. Discrete Methods* **1**, 1–7. (Cited in §7.1.)

**R. Grone, C. R. Johnson, E. M. Sá, & H. Wolkowicz (1984).**
Positive definite completions of partial Hermitian matrices. *Linear Algebra Appl.* **58**, 109–124. (Cited in §2.4.3, 7.8.)

**L. Guibas, J. Pach, & M. Sharir (1994).**
Sphere-of-influence graphs in higher dimensions. In *Intuitive Geometry* (K. Böröczky & G. F. Tóth, eds.), *Colloq. Math. Sci. János Bolyai* **63**, North–Holland, Amsterdam; pp. 131–137. (Cited in §7.11.)

**D. R. Guichard (1998).**
Competition graphs of hamiltonian digraphs. *SIAM J. Discrete Methods* **11**, 128–134. (Cited in §7.1.)

**D. Gusfield (1991).**
Efficient algorithms for inferring evolutionary trees. *Networks* **21**, 19–28. (Cited in §2.4.1.)

**M. Gutierrez & L. Oubiña (1995).**
Minimum proper interval graphs. *Discrete Math.* **142**, 77–85. (Cited in §3.3.)

**M. Gutierrez & L. Oubiña (1996).**
Metric characterizations of proper interval graphs and tree-clique graphs. *J. Graph Theory* **21**, 199–205. (Cited in §3.3, 7.5.)

**A. Gyárfás & J. Lehel (1988).**
On-line and first fit colorings of graphs. *J. Graph Theory* **12**, 217–227. (Cited in §3.4.)

**A. Gyárfás & D. B. West (1995).**
Multitrack interval graphs. *Congr. Numer.* **109**, 109–116. (Cited in §7.1.)

**A. Hajnal & J. Surányi (1958).**
Über die Auflösung von Graphen in vollständige Teilgraphen. *Ann. Univ. Sci. Budapest. Eötvös. Sect. Math.* **1**, 113–121. (Cited in beginning of §2, §7.7.)

**G. Hajós (1957).**
Über eine Art von Graphen. *Internat. Math. Nachr.* **11**, Problem 65. (Cited in beginning of §3.)

**R. Halin (1982).**
Some remarks on interval graphs. *Combinatorica* **2**, 297–304. (Cited in §7.7.)

**R. Halin (1984).**
On the representation of triangulated graphs in trees. *Europ. J. Combin.* **5**, 23–28. (Cited in §7.7.)

**R. Hamelink (1968).**
A partial characterization of clique graphs. *J. Combin. Theory* **5**, 192–107. (Cited in §7.5.)

**P. L. Hammer, F. Maffray, & M. Preissmann (1989).**
A characterization of chordal bipartite graphs. Rutcor Research Report **16-89**, Rutgers University, New Brunswick, NJ. (Cited in §7.3, 7.12.)

**P. L. Hammer, U. N. Peled, & X. Sun (1990).**
Difference graphs. *Discrete Appl. Math.* **28**, 35–44. (Cited in §5.3.)

**F. Harary (1963).**
A characterization of block-graphs. *Canad. Math. Bull.* **6**, 1–6. (Cited in §7.10.)

**F. Harary, M. S. Jacobson, M. J. Lipman, & F. R. McMorris (1993).** Abstract sphere-of-influence graphs. *Math. Comput. Modelling* **17**, 77–83. (Cited in §7.11.)

**F. Harary, M. S. Jacobson, M. J. Lipman, & F. R. McMorris (1994).** Sphere of influence graphs defined on a prescribed graph. *J. Combin. Inform. System Sci.* **19**, 5–10. (Cited in §7.11.)

**F. Harary, M. S. Jacobson, M. J. Lipman, & F. R. McMorris (1995).** Trees that are sphere of influence graphs. *Appl. Math. Lett.* **8**, 89–93. (Cited in §7.11.)

**F. Harary & J. A. Kabell (1984).**
Infinite-interval graphs. In *Calcutta Mathematical Society. Diamond-cum-Platinum Jubilee Commemoration Volume* (1908–1983), Part I, Calcutta Math. Soc., Calcutta; pp. 27–31. (Cited in §3.1.)

**F. Harary, J. A. Kabell, & F. R. McMorris (1982).**
Bipartite intersection graphs. *Comment. Math. Univ. Carolin.* **23**, 739–745. (Cited in §7.2.)

**F. Harary, J. A. Kabell, & F. R. McMorris (1990).**
Interval acyclic digraphs. *Ars Combin.* **29A**, 59–64. (Cited in §7.2.)

**F. Harary, J. A. Kabell, & F. R. McMorris (1992).**
Subtree acyclic digraphs. *Ars Combin.* **34**, 93–95. (Cited in §7.2.)

**F. Harary & T. A. McKee (1994).**
The square of a chordal graph. *Discrete Math.* **128**, 165–172. (Cited in §6.2.)

**F. Harary & I. C. Ross (1960).**
The square of a tree. *Bell System Tech. J.* **39**, 641–647. (Cited in §4.1.1.)

**D. R. Hare & W. McCuaig (1993).**
The connectivity of the block-intersection graphs of designs. *Des. Codes Cryptogr.* **3**, 5–8. (Cited in §7.8.)

**I. Hartman, I. Newman, & R. Ziv (1991).**
On grid intersection graphs. *Discrete Math.* **87**, 41–52. (Cited in §7.1.)

**R. B. Hayward (1985).**
Weakly triangulated graphs. *J. Combin. Theory Ser. B* **39**, 200–209. (Cited in §7.3.)

**R. B. Hayward (1996).**
Generating weakly triangulated graphs. *J. Graph Theory* **21**, 67–69. (Cited in §7.3.)

**R. B. Hayward, C. T. Hoàng, & F. Maffray (1989).**
Optimizing weakly triangulated graphs. *Graphs Combin.* **5**, 339–349. [Erratum, *ibid.*, **6** (1990), 33–35.] (Cited in §7.3.)

**B. Hedman (1984).**
Clique graphs of time graphs. *J. Combin. Theory Ser. B* **37**, 270–278. (Cited in §3.3.)

**K. A. S. Hefner, J. F. Jones, S.-R. Kim, J. R. Lundgren, & F. S. Roberts (1991).** $(i, j)$ competition graphs. *Discrete Appl. Math.* **32**, 241–262. (Cited in §4.2.)

**P. Hell & J. Huang (1995).**
Lexicographic orientations and representation algorithms for comparability graphs, proper circular arc graphs, and proper interval graphs. *J. Graph Theory* **20**, 361–374. (Cited in §3.3, 7.1, 7.6.)

**P. Hell & J. Huang (1997).**
Two remarks on circular arc graphs. *Graphs Combin.* **13**, 65–72. (Cited in §7.1.)

**R. L. Hemminger & L. W. Beineke (1978).**
Line graphs and line digraphs. In *Selected Topics in Graph Theory* (L. W. Beineke & R. J. Wilson, eds.), Academic Press, London; pp. 271–305. (Cited in §1.5.)

**P. B. Henderson & Y. Zalcstein (1977).**
A graph-theoretic characterization of the $PV_{chunk}$ class of synchronizing primitives. *SIAM J. Comput.* **6**, 88–108. (Cited in §5.1, 5.4.)

**P. Hliněný & A. Kuběna (1995).**
A note on intersection dimensions of graph classes. *Comment. Math. Univ. Carolin.* **36**, 255-261. (Cited in §7.1.)

**C.-W. Ho & R. C. T. Lee (1989).**
Counting clique trees and computing perfect elimination schemes in parallel. *Inform. Process. Lett.* **31**, 61–68. (Cited in §2.1.)

**A. J. Hoffman, A. W. J. Kolen, & M. Sakarovitch (1985).**
Totally balanced and greedy matrices. *SIAM J. Alg. Discrete Methods* **6**, 721–730. (Cited in §7.3, 7.12.)

**T. S. Holm & K. P. Bogart (to appear).**
Trees are tolerance sphere-of-influence graphs. (Cited in §6.3, 7.11.)

**W.-L. Hsu (1993).**
A simple test for interval graphs. In *Graph-Theoretic Concepts for Computer Science* (E. W. Mayr, ed.), [*Lecture Notes in Computer Science* **657**], Springer, Berlin; pp. 11–16. (Cited in §3.1.)

**W.-L. Hsu (1995).**
$O(mn)$ algorithms for the recognition and isomorphism problems on circular-arc graphs. *SIAM J. Comput.* **24**, 411–439. (Cited in §7.1.)

**W.-L. Hsu & T. H. Ma (1991).**
Substitution decomposition on chordal graphs and applications. In *ISA '91 Algorithms* (W.-L. Hsu & R. C. T. Lee, eds.), [*Lecture Notes in Computer Science* **557**], Springer, Berlin; pp. 52–60. (Cited in §2.1, 3.1.)

**L. Hubert (1974).**
Some applications of graph theory and related non-metric techniques to problems of approximate seriation: The case of symmetric proximity measures. *British J. Math. Statist. Psychology* **27**, 133–153. (Cited in §3.4.2.)

**R. M. Indury & A. A. Schaeffer (1993).**
Triangulating three-colored graphs. *SIAM J. Discrete Math.* **6**, 289–293. (Cited in §2.4.1.)

**G. Isaak, S.-K. Kim, T. A. McKee, F. R. McMorris, & F. S. Roberts (1992).** 2-competition graphs. *SIAM J. Discrete Math.* **5**, 524–538. (Cited in §6.1.)

**Z. Jackowski (1992).**
A new characterization of proper interval graphs. *Discrete Math.* **105**, 103–109. (Cited in §3.3.)

**M. S. Jacobson (1992).**
On the $p$-edge clique cover number of complete bipartite graphs. *SIAM J. Discrete Math.* **5**, 539–544. (Cited in §6.1.)

**M. S. Jacobson, A. E. Kézdy, & D. B. West (1995).**
The 2-intersection number of paths and bounded-degree trees. *J. Graph Theory* **19**, 461–469. (Cited in §6.1.)

**M. S. Jacobson, J. Lehel, & L. M. Lesniak (1993).**
$\phi$-threshold and $\phi$-tolerance chain graphs. *Discrete Appl. Math.* **44**, 191–203. (Cited in §6.3.)

**M. S. Jacobson, M. J. Lipman, & F. R. McMorris (1995).**
Trees that are sphere-of-influence graphs. *Appl. Math. Letters*, **8**, 89–93. (Cited in §7.11.)

**M. S. Jacobson & F. R. McMorris (1991).**
Sum-tolerance proper interval graphs are precisely sum-tolerance unit interval graphs. *J. Combin. Inform. System Sci.*, **16**, 25–28. (Cited in §6.3.)

**M. S. Jacobson, F. R. McMorris, & H. M. Mulder (1991).**
An introduction to tolerance intersection graphs. In *Graph Theory, Combinatorics and Applications* (Y. Alavi, G. Chartrand, O. R. Oellermann, & A. J. Schwenk, eds.) Wiley Interscience, New York; Vol. 2, pp. 705–723. (Cited in §6.3.)

**M. S. Jacobson, F. R. McMorris, & E. R. Scheinerman (1991).**
General results on tolerance intersection graphs. *J. Graph Theory* **15**, 573–577. (Cited in §6.1, 6.3.)

**R. Jamison (to appear).**
Powers of block graphs are chordal. (Cited in §7.10.)

**S. Janson & J. Kratochvíl (1992).**
Thresholds for classes of intersection graphs. *Discrete Math.* **108**, 307–326. (Cited in §7.8.)

**J. W. Jaromczyk & G. T. Toussaint (1992).**
Relative neighborhood graphs and their relatives. *IEEE Proc.* **80**, 1502–1517. (Cited in §7.11.)

**C. R. Johnson (1990).**
Matrix completion problems: A survey. In *Matrix Theory and Applications* (C. R. Johnson, ed.), [*Proceedings of Symposia in Applied Mathematics* **40**], American Mathematical Society, Providence, RI; pp. 171–198. (Cited in §2.4.3.)

**C. R. Johnson, C. A. Jones, & B. K. Kroschel (1995).**
The Euclidean distance completion problem: Cycle completability. *Linear and Multilinear Algebra* **39**, 195–207. (Cited in §2.4.3.)

**C. R. Johnson & J. Miller (1997).**
Rank decomposition under combinatorial constraints. *Linear Algebra Appl.* **251**, 97–104. (Cited in §7.3.)

**C. R. Johnson & G. T. Whitney (1991).**
Minimum rank completions. *Linear and Multilinear Algebra* **28**, 271–273. (Cited in §7.3.)

**D. Joseph, J. Meidanis, & P. Tiwari (1992).**
Determining DNA sequence similarity using maximum independent set algorithms for interval graphs. In *Algorithm Theory — SWAT '92* (O. Nurmi & E. Ukkonen, eds.), [*Lecture Notes in Computer Science* **621**], Springer, Berlin; pp. 326–337. (Cited in §7.1.)

**J. R. Jungck, G. Dick, & A. G. Dick (1982).**
Computer-assisted sequencing, interval graphs, and molecular evolution. *BioSystems* **15**, 259–273. (Cited in §3.4.1.)

**S. K. Kannan & T. J. Warnow (1992).**
Triangulating 3-colored graphs. *SIAM J. Discrete Math.* **5**, 249–258. (Cited in §2.4.1.)

**S. Kannan & T. Warnow (1994).**
Inferring evolutionary history from DNA sequences. *SIAM J. Comput.* **23**, 713–737. (Cited in §2.4.1.)

**L. L. Kelleher & M. B. Cozzens (1990).**
Coloring interval graphs with First-Fit. *Discrete Math.* **86**, 101–116. (Cited in §7.9.)

**D. G. Kendall (1969).**
Incidence matrices, interval graphs, and seriation in archaeology. *Pacific J. Math.* **28**, 565–570. (Cited in §3.4.3.)

**A. D. Kézdy & G. Kubicki (1997).**
$K_{12}$ is not a closed sphere-of-influence graph. *Bolyai Soc. Math. Stud.* **6**, 383–397. (Cited in §7.11.)

**H. J. Khamis (1996).**
Application of the multigraph representation of hierarchical log-linear models. In *Categorical Variables in Developmental Research: Methods of Analysis* (A. von Eye & C. C. Clogg, eds.), Academic Press, New York; pp. 215–229. (Cited in §2.4.4.)

**H. J. Khamis & T. A. McKee (1997).**
Chordal graph models of contingency tables. *Comput. Math. Appl.* **34**, 89–97. (Cited in §2.4.4.)

**H. A. Kierstead (1991).**
A polynomial time approximation algorithm for dynamic storage allocation. *Discrete Math.* **88**, 231–237. (Cited in §3.4.)

**H. A. Kierstead & J. Qin (1995).**
Coloring interval graphs with First-Fit. *Discrete Math.* **144**, 47–57. (Cited in §3.4.)

**S.-R. Kim (1993).**
The competition number and its variants. In *Quo Vadis, Graph Theory?* (J. Gimbel, J. W. Kennedy, & L. V. Quintas, eds.), [*Ann. Discrete Math.* **55**], North–Holland, Amsterdam; pp. 313–326. (Cited in §4.2.)

**S.-R. Kim, T. A. McKee, F. R. McMorris, & F. S. Roberts (1993).**
*p*-competition numbers. *Discrete Appl. Math.* **46**, 87–92. (Cited in §6.1.)

**S.-R. Kim, T. A. McKee, F. R. McMorris, & F. S. Roberts (1995).**
*p*-competition graphs. *Linear Algebra Appl.* **217**, 167–178. (Cited in §6.1.)

**S.-R. Kim & F. S. Roberts (1990).**
On Opsut's conjecture about the competition number. *Congr. Numer.* **71**, 173–176. (Cited in §4.2.)

**S.-R. Kim & F. S. Roberts (1997).**
Competition numbers of graphs with a small number of triangles. *Discrete Appl. Math.* **78**, 153–162. (Cited in §4.2.)

**D. J. Klein (1982).**
Treediagonal matrices and their inverses. *Linear Algebra Appl.* **42**, 109–117. (Cited in §2.4.3.)

**P. N. Klein (1996).**
Efficient parallel algorithms for chordal graphs. *SIAM J. Comput.* **25**, 797–827. (Cited in §2.4.2.)

**T. Kloks (1994).**
*Treewidth: Computations and Approximations* [*Lecture Notes in Computer Science* **842**]. Springer, Berlin. (Cited in Preface, §7.1)

**T. Kloks, H. Bodlaender, H. Müller, & D. Kratsch (1993).**
Computing treewidth and minimum fill-in: All you need are the minimal separators. In *Algorithms—ESA '93* (T. Lengauer, ed.) [*Lecture Notes in Computer Science* **726**], Springer, Berlin; pp. 260–271. (Cited in §2.4.2.)

**T. Kloks & D. Kratsch (1994).**
Finding all minimal separators of a graph. In *STACS 94* (P. Enjalbert, E. W. Mayr, & K. W. Wagner, eds.) [*Lecture Notes in Computer Science* **775**], Springer, Berlin; pp. 759–768. (Cited in §2.4.2.)

**T. Kloks & D. Kratsch (1995).**
Computing a perfect edge without vertex elimination ordering of a chordal bipartite graph. *Inform. Process. Lett.* **55**, 11–16. (Cited in §7.3.)

**T. Kloks, D. Kratsch, & C. K. Wong (1996).**
Minimum fill-in on circle and circular-arc graphs. In *Automata, Languages and Programming* (F. Meyer auf der Heide & B. Monien, eds.) [*Lecture Notes in Computer Science* **1099**], Springer, Berlin; pp. 256–267. (Cited in §7.1, 7.4.)

**G. J. Koop (1986).**
Cyclic scheduling of offweekends. *Oper. Res. Lett.* **4**, 259–263. (Cited in §5.4.)

**L. T. Kou, L. J. Stockmeyer, & C. K. Wong (1978).**
Covering graphs by cliques with regard to keyword conflicts and intersection graphs. *Comm. ACM* **21**, 135–139. (Cited in §1.3.)

**J. Kratochvíl (1991a).**
String graphs. I: The number of critical nonstring graphs is infinite. *J. Combin. Theory Ser. B* **52**, 53–66. (Cited in §7.1.)

**J. Kratochvíl (1991b).**
String graphs. II. Recognizing string graphs is NP-hard. *J. Combin. Theory Ser. B* **52**, 67–78. (Cited in §7.1.)

**J. Kratochvíl & J. Matoušek (1994).**
Intersection graphs of segments. *J. Combin. Theory Ser. B* **62**, 289–315. (Cited in §7.1.)

**J. Kratochvíl & Z. Tuza (1994).**
Intersection dimensions of graph classes. *Graphs Combin.* **10**, 159–168. (Cited in §7.1.)

**T. M. Kratzke & D. B. West (1993).**
The total interval number of a graph. I: Fundamental classes. *Discrete Math.* **118**, 145–156. (Cited in §7.1.)

**T. M. Kratzke & D. B. West (1996).**
The total interval number of a graph. II: Trees and complexity. *SIAM J. Discrete Math.* **9**, 339–348. (Cited in §7.1.)

**J. Krausz (1943).**
Démonstration nouvelle d'une théorème de Whitney sur les réseaux. *Mat. Fiz. Lapok* **50**, 75–85. (Cited in §1.5.)

**N. Kumar & N. Deo (1994).**
Multidimensional interval graphs. *Congr. Numer.* **102**, 45–56. (Cited in §7.1.)

**P. S. Kumar & C. E. Veni Madhavan (1989).**
A new class of separators and planarity of chordal graphs. In *Foundations of Software Technology and Theoretical Computer Science* (G. Goos & J. Hartmanis, eds.) [*Lecture Notes in Computer Science* **405**], Springer, Berlin; pp. 30–43. (Cited in §2.2.)

**L. Langley (1994).**
A note on bipartite interval tolerance graphs. *Congr. Numer.* **102**, 191–192. (Cited in §7.6.)

**L. Langley, J. R. Lundgren, & S. K. Merz (1995).**
The competition graphs of interval digraphs. *Congr. Numer.* **107**, 37–40. (Cited in §7.2.)

**R. Laskar & D. Shier (1980).**
On chordal graphs. *Congr. Numer.* **29**, 579–588. (Cited in §7.10.)

**R. Laskar & D. Shier (1983).**
On powers and centers of chordal graphs. *Discrete Appl. Math.* **6**, 139–147. (Cited in §7.10.)

**S. L. Lauritzen (1996).**
*Graphical Models.* [*Oxford Statistical Sciences Series* **17**] Clarendon Press, Oxford. (Cited in §2.4.4.)

**S. L. Lauritzen & D. J. Spiegelhalter (1988).**
Local computations with probabilities on graphical structures and their applications to expert systems. *J. R. Statist. Soc. Ser. B* **50**, 157–224. (Cited in §2.4.4.)

**J. Lehel (1983).**
Helly hypergraphs and abstract interval structures. *Ars Combin.* **16A**, 239–253. (Cited in §2.3, 3.2.)

**J. Lehel (1985).**
A characterization of totally balanced hypergraphs. *Discrete Math.*
**57**, 59–65. (Cited in §2.3.)

**P. G. H. Lehot (1974).**
An optimal algorithm to detect a line graph and output its root graph.
*J. Assoc. Comput. Mach.* **21**, 569–575. (Cited in §1.5.)

**R. Leibowitz (1978).**
*Interval Counts and Threshold Graphs.* Ph.D. Thesis, Rutgers University, New Brunswick, NJ. (Cited in §5.2, 5.4.)

**R. Leibowitz, S. F. Assman, & G. W. Peck (1982).**
Interval counts of interval graphs. *SIAM J. Alg. Discrete Methods* **3**,
485–494. (Cited in §3.3.)

**C. G. Lekkerkerker & J. C. Boland (1962).**
Representation of a finite graph by a set of intervals on the real line.
*Fund. Math.* **51**, 45–64. (Cited in §3.1.)

**M. Lewin (1983).**
On intersection multigraphs of hypergraphs. *J. Combin. Theory Ser.*
*B* **34**, 229–232. (Cited in §2.1.)

**K.-W. Lih (1993).**
Rank inequalities for chordal graphs. *Discrete Math.* **113**, 125–130.
(Cited in §2.1.)

**C. K. Lim (1978).**
On supercompact graphs. *J. Graph Theory* **2**, 349–355. (Cited in
§1.3.)

**C. K. Lim & Y. H. Peng (1991).**
Uniquely pseudointersectable graphs. *Ars Combin.* **32**, 3–11. (Cited
in §1.3.)

**I.-J. Lin, T. A. McKee, & D. B. West (to appear).**
Leafage of chordal graphs. *Discuss. Math. Graph Theory.* (Cited in
§2.1.)

**I.-J. Lin & D. B. West (1995).**
Interval digraphs that are indifference digraphs. In *Graph Theory,*
*Combinatorics, and Applications* (Y. Alavi & A. Schwenk, eds.) Wiley-
Interscience, New York; Vol. 2, pp. 751–765. (Cited in §7.2.)

**M. J. Lipman (1992).**
Integer realizations of sphere-of-influence graphs. *Congr. Numer.* **91**, 63–70. (Cited in §7.11.)

**M. J. Lipman (1996).**
Maximum tolerance sphere-of-influence graphs. *Congr. Numer.* **121**, 195–203. (Cited in §7.11.)

**W. Lipski (1983).**
The consecutive retrieval property, interval graphs and related topics— A survey. In *Data Base File Organization, Theory and Applications of the Consecutive Retrieval Property* (S. P. Ghosh, Y. Kambayashi, & W. Lipski, eds.), Academic Press, New York; pp. 17–54. (Cited in §3.4.3.)

**A. Lubiw (1982).**
$\Gamma$-*Free Matrices.* M.S. Thesis, University of Waterloo. (Cited in §7.3, 7.10.)

**A. Lubiw (1987).**
Doubly lexical orderings of matrices. *SIAM J. Comput.* **16**, 854–879. (Cited in §7.3, 7.12.)

**R. D. Luce (1956).**
Semiorders and a theory of utility discrimination. *Econometrica* **24**, 178–191. (Cited in §3.4.2.)

**J. K. Luedeman (1987).**
Intersection graphs of semigroups II: Quasi-ideals and bi-ideals. *Congr. Numer.* **59**, 205–209. (Cited in §7.8.)

**J. K. Luedeman & F. R. McMorris (1986).**
Intersection graphs of semigroups. *Congr. Numer.* **55**, 31–37. (Cited in §7.8.)

**J. R. Lundgren (1989).**
Food webs, competition graphs, competition-common enemy graphs, and niche graphs. In *Applications of Combinatorics and Graph Theory to the Biological and Social Sciences* (F. S. Roberts, ed.), Springer, New York; pp. 221–243. (Cited in §4.2, 4.3.)

**J. R. Lundgren & J. S. Maybee (1983a).**
A characterization of graphs of competition number $m$. *Discrete Appl. Math.* **6**, 319–322. (Cited in §4.2.)

**J. R. Lundgren & J. S. Maybee (1983b).**
A characterization of upper bound graphs. *Congr. Numer.* **40**, 189–193. (Cited in §4.4.)

**J. R. Lundgren & J. S. Maybee (1984).**
Food webs with interval competition graphs. In *Graphs and Applications: Proceedings of the First Colorado Symposium on Graph Theory* (F. Harary & J. S. Maybee, eds.), Wiley, New York; pp. 245–256. (Cited in §4.3.)

**J. R. Lundgren, J. S. Maybee, & F. R. McMorris (1988).**
Two-graph inversion of competition graphs and bound graphs. *Congr. Numer.* **67**, 136–144. (Cited in §4.4.)

**J. R. Lundgren, P. A. McKenna, L. Langley, S. K. Merz, & C. W. Rasmussen (1997).** The $p$-competition graphs of strongly-connected and hamiltonian digraphs. *Ars Combin.* **47**, 161–172. (Cited in §6.1.)

**J. R. Lundgren, P. A. McKenna, S. K. Merz, & C. W. Rasmussen (1995).** Interval $p$-neighborhood graphs. *Congr. Numer.* **108**, 3–10. (Cited in §6.1.)

**J. R. Lundgren, P. A. McKenna, S. K. Merz, & C. W. Rasmussen (to appear).** The $p$-competition graphs of symmetric digraphs and $p$-neighborhood graphs. (Cited in §6.1.)

**J. R. Lundgren & S. K. Merz (1994).**
Elimination ordering characterizations of digraphs with interval and chordal competition graphs. *Congr. Numer.* **103**, 55–64. (Cited in §4.3.)

**J. R. Lundgren, S. Merz, J. S. Maybee, & C. W. Rasmussen (1995).**
A characterization of graphs with interval two-step graphs. *Linear Algebra Appl.* **217**, 203–223. (Cited in §4.3.)

**J. R. Lundgren, S. K. Merz, & C. W. Rasmussen (1993).**
A characterization of graphs with interval squares. *Congr. Numer.* **98**, 132–142. (Cited in §4.3, 7.10.)

**T. H. Ma & J. P. Spinrad (1991).**
Cycle-free partial orders and chordal comparability graphs. *Order* **8**, 49–61. (Cited in §7.6.)

**S. Ma, W. D. Wallis, & J. Wu (1989).**
Optimization problems on quasi-threshold graphs. *J. Combin. Inform. System Sci.* **14**, 105–110. (Cited in §7.9.)

**S. Ma & J. Wu (1990).**
Characterizing strongly chordal graphs by using minimal relative separators. In *Combinatorial Designs and Applications* (W. D. Wallis, H. Shen, W. Wei, & L. Zhu, eds.), [*Lecture Notes in Pure and Applied Mathematics* **126**] Marcel Dekker, New York; pp. 87–95. (Cited in §7.12.)

**H. Maehara (1984a).**
A digraph represented by a family of boxes or spheres. *J. Graph Theory* **8**, 431–439. (Cited in §7.2.)

**H. Maehara (1984b).**
Space graphs and sphericity. *Discrete Appl. Math.* **7**, 55–64. (Cited in §7.1.)

**H. Maehara (1990).**
On the intersection graphs of random arcs on a circle. In *Random Graphs '87* (M. Karoński, J. Jaworski, & A. Ruciński, eds.), Wiley, Chichester; pp. 159–173. (Cited in §7.8.)

**H. Maehara (1991).**
The intersection graph of random sets. *Discrete Math.* **87**, 97–104. (Cited in §7.8.)

**N. V. R. Mahadev & U. N. Peled (1995).**
*Threshold Graphs and Related Topics.* [*Ann. of Discrete Math.* **56**], North–Holland, Amsterdam. (Cited in Preface and throughout Chapter 5.)

**N. V. R. Mahadev & T.-M. Wang (1997).**
A characterization of hereditary UIM graphs. *Congr. Numer.* **126**, 183–191. (Cited in §1.3.)

**N. V. R. Mahadev & T.-M. Wang (to appear).**
On uniquely intersectable graphs. (Cited in §1.3.)

**F. Maire (1993).**
A characterization of intersection graphs of the maximal rectangles of a polyomino. *Discrete Math.* **120**, 211–214. (Cited in §7.1.)

**G. Major & F. R. McMorris (1990).**
p-edge clique coverings of graphs. *Congr. Numer.* **79**, 143–145. (Cited in §6.1.)

**M. V. Marathe, H. Breu, H. B. Hunt, S. S. Ravi, & D. J. Rosen-krantz (1995).** Simple heuristics for unit disk graphs. *Networks* **25**, 59–68. (Cited in §7.1.)

**M. V. Marathe, H. B. Hunt, & S. S. Ravi (1996).**
Efficient approximation algorithms for domatic partition and on-line coloring of circular arc graphs. *Discrete Appl. Math.* **64**, 135–149. (Cited in §3.4.)

**E. (Szpilrajn-) Marczewski (1945).**
Sur deux propriétés des classes d'ensembles. *Fund. Math.* **33**, 303–307. (Cited in §1.1.)

**M. L. N. McAllister (1988).**
Fuzzy intersection graphs. *Comput. Math. Appl.* **15**, 871–886. (Cited in §7.8.)

**K. McAvaney, J. Robertson, & D. DeTemple (1993).**
A characterization and hereditary properties for partition graphs. *Discrete Math.* **113**, 131–142. (Cited in §7.8.)

**R. M. McConnell & J. P. Spinrad (1994).**
Linear-time modular decomposition and efficient transitive orientation of comparability graphs. In *Proceedings of the Fifth Annual ACM-SIAM Symposium on Discrete Algorithms*, ACM, New York; pp. 536–545. (Cited in §7.4.)

**T. A. McKee (1978).**
Forbidden subgraphs in terms of forbidden quantifiers. *Notre Dame J. Formal Logic* **19**, 186–188. (Cited in §1.2.)

**T. A. McKee (1987).**
Bipartite analogs of graph theory. *Congr. Numer.* **60**, 261–268. (Cited in §7.2.)

**T. A. McKee (1989).**
Upper bound multigraphs for posets. *Order* **6**, 265–275. (Cited in §6.2.)

**T. A. McKee (1990a).**
Neighborhood and self-dual (multi)graphs. *J. Combin. Math. Combin. Comput.* **8**, 173–180. (Cited in §6.2.)

**T. A. McKee (1990b).**
Interval competition multigraphs of food webs. *Congr. Numer.* **71**, 197–204. (Cited in §6.2.)

**T. A. McKee (1990c).**
Intersection graphs and cographs. *Congr. Numer.* **78**, 223–230. (Cited in §7.9.)

**T. A. McKee (1991a).**
Foundations of intersection graph theory. *Utilitas Math.* **40**, 77–86. (Cited in §1.5, 6.1.)

**T. A. McKee (1991b).**
Chordal and interval multigraphs. In *Graph Theory, Combinatorics and Applications* (Y. Alavi, G. Chartrand, O. R. Ollermann, & A. J. Schwenk, eds.) Wiley-Interscience, New York; Vol. 2, pp. 841–848. (Cited in §6.2.)

**T. A. McKee (1991c).**
Clique multigraphs. In *Graph Theory, Combinatorics, Algorithms and Applications* (Y. Alavi, F. R. K. Chung, R. L. Graham, & D. F. Hsu, eds.) Society for Industrial and Applied Mathematics, Philadelphia; pp. 371–379. (Cited in §6.2.)

**T. A. McKee (1991d).**
Intersection properties of graphs. *Discrete Math.* **89**, 253–260. (Cited in §1.5.)

**T. A. McKee (1992).**
Subtree catch graphs. *Congr. Numer.* **90**, 231–238. (Cited in §7.2.)

**T. A. McKee (1993).**
How chordal graphs work. *Bull. Inst. Combin. Appl.* **9**, 27–39. (Cited in §2.1, 2.4.4.)

**T. A. McKee (1994).**
Clique pseudographs and pseudo duals. *Ars Combin.* **38**, 161–173. (Cited in §6.2, 7.5)

**T. A. McKee (1995a).**
Niche space, multigraphs, and the Helly condition. *Math. Comput. Modelling* **22**, 1–8. (Cited in §4.3, 6.2.)

**T. A. McKee (1995b).**
A survey of connection graphs. In *Graph Theory, Combinatorics, and Applications* (Y. Alavi & A. Schwenk, eds.), Wiley-Interscience, New York; Vol. 2, pp. 767–776. (Cited in §7.6.)

**T. A. McKee (1998).**
F-Free interval graphs. *Utilitas Math.*, **53**, 147–158. (Cited in §7.9.)

**T. A. McKee (to appear(a)).**
An inequality characterizing chordal graphs. *Ars Combin.*, (Cited in §7.9.)

**T. A. McKee (to appear(b)).**
Strong clique trees and strongly chordal graphs. (Cited in §7.12.)

**T. A. McKee & H. J. Khamis (1996).**
Multigraph representations of hierarchical loglinear models. *J. Statist. Plann. Inference* **53**, 63–74. (Cited in §2.4.4)

**T. A. McKee & F. R. McMorris (1992).**
Comparability multigraphs. *Congr. Numer.* **89**, 33–38. (Cited in §7.6.)

**T. A. McKee & E. R. Scheinerman (1993).**
On the chordality of a graph. *J. Graph Theory*, **17**, 221–232. (Cited in §7.1.)

**F. R. McMorris (1977).**
On the compatibility of binary qualitative taxonomic characters. *Bull. Math. Biology* **39**, 133–138. (Cited in §2.4.1.)

**F. R. McMorris & C. A. Meacham (1983).**
Partition intersection graphs. *Ars Combin.* **16B**, 135–138. (Cited in §2.4.1.)

**F. R. McMorris & H. M. Mulder (1996).**
Subpath acyclic digraphs. *Discrete Math.* **154**, 189–201. (Cited in §7.2.)

**F. R. McMorris & G. T. Myers (1983).**
Some uniqueness results for upper bound graphs. *Discrete Math.* **44**, 321–323. (Cited in §4.4.)

**F. R. McMorris & E. R. Scheinerman (1991).**
Connectivity threshold for random chordal graphs. *Graphs. Combin.* **7**, 177–181. (Cited in §7.8.)

**F. R. McMorris & D. R. Shier (1983).**
Representing chordal graphs on $K_{1,n}$. *Comment. Math. Univ. Carolin.* **24**, 489–494. (Cited in §2.5.)

**F. R. McMorris & C. Wang (1996).**
Modular intersection graphs. *Graphs Combin.* **12**, 267–281. (Cited in §6.1.)

**F. R. McMorris & C. Wang (to appear).**
Sphere-of-attraction graphs. *SIAM J. Discrete Math.*, (Cited in §7.11.)

**F. R. McMorris, C. Wang, & P. Zhang (to appear).**
On probe interval graphs. *Discrete Appl. Math.* (Cited in §3.4.1.)

**F. R. McMorris, T. J. Warnow, & T. Wimer (1994).**
Triangulating vertex-colored graphs. *SIAM J. Discrete Math.* **7**, 296–306. (Cited in §2.4.1.)

**F. R. McMorris & T. Zaslavsky (1982).**
Bound graphs of a partially ordered set. *J. Combin. Inform. System Sci.* **7**, 134–138. (Cited in §4.4.)

**C. A. Meacham (1983).**
Theoretical and computational considerations of the compatibility of qualitative taxonomic characters. In *Numerical Taxonomy: Proceedings of a NATO Advanced Study Institute* (J. Felsenstein, ed.), *NATO Advanced Study Institute Ser. G* **1**, pp. 304–314. (Cited in §2.4.1.)

**T. S. Michael & T. Quint (1994).**
Sphere of influence graphs: A survey. *Congr. Numer.* **105**, 153–160. (Cited in §7.11.)

**T. S. Michael & T. Quint (to appear).**
Sphere of influence graphs in the plane. (Cited in §7.11.)

**B. G. Mirkin & S. N. Rodin (1984).**
*Graphs and Genes.* Springer, Berlin. (Cited in §3.4.1.)

**R. H. Möhring (1996).**
Triangulating graphs without asteroidal triples. *Discrete Appl. Math.* **64**, 281–287. (Cited in §7.6.)

**C. L. Monma, B. Reed, & W. T. Trotter (1988).**
Threshold tolerance graphs. *J. Graph Theory* **12**, 343–362. (Cited in §6.3.)

**C. L. Monma & V. K. Wei (1986).**
Intersection graphs of paths in a tree. *J. Combin. Theory Ser. B* **41**, 141–181. (Cited in §7.1, 7.2.)

**T. B. Moorhouse (1994).**
*Characterizing Hereditary Graph Classes by Subgraph Intersections.* M.S. thesis, University of Toronto. [Technical Report **290/94**, Dept. Comput. Sci.] (Cited in §1.2.)

**T. B. Moorhouse (to appear(a)).**
Characterizing hereditary graph classes as the intersection graphs of subgraphs of graphs. (Cited in §1.2.)

**T. B. Moorhouse (to appear(b)).**
Completeness for intersection classes. (Cited in §1.2.)

**A. Mukhopadhyay (1967).**
The square root of a graph. *J. Combin. Theory* **2**, 290–295. (Cited in §4.1.1.)

**H. Müller (1997).**
Recognizing interval digraphs and interval bigraphs in polynomial time. *Discrete Appl. Math.*, **78**, 189–205. (Cited in §7.2, 7.3.)

**G. T. Myers (1982).**
*Upper Bound Graphs of Partially Ordered Sets.* Ph.D. thesis, Bowling Green State University, Bowling Green, OH. (Cited in §7.9.)

**D. Q. Naiman & H. P. Wynn (1992).**
Inclusion-exclusion-Bonferroni identities and inequalities for discrete tube-like problems via Euler characteristics. *Ann. Statist.* **20**, 43–76. (Cited in §2.3.)

**W. Naji (1985).**
Reconnaissance des graphes de cordes. *Discrete Math.* **54**, 327–337. (Cited in §7.4.)

**G. Narasimhan & R. Manber (1992).**
Stability number and chromatic number of tolerance graphs. *Discrete Appl. Math.* **36**, 47–56. (Cited in §6.3.)

**W. M. Nawijn (1991).**
On a random interval graph and the maximum throughput rate in the system $GI/G/1/0$. *Adv. in Appl. Probab.* **23**, 945–956. (Cited in §7.8.)

**R. E. Neapolitan (1990).**
*Probabilistic Reasoning in Expert Systems.* Wiley, New York. (Cited in §2.4.4.)

**V. Nicholson (1995).**
Applying interval graphs to computing a protein model. In *Graph Theory, Combinatorics, and Applications* (Y. Alavi & A. Schwenk, eds.), Wiley-Interscience, New York; Vol. 2, pp. 833–838. (Cited in §3.4.1.)

**R. M. Odom & C. W. Rasmussen (1995).**
Conditional completion algorithms for classes of chordal graphs. *Congr. Numer.* **109**, 97–108. (Cited in §7.8.)

**W. F. Ogden & F. S. Roberts (1970).**
Intersection graphs of families of convex sets with distinguished points. In *Combinatorial Structures and Their Applications* (R. Guy, H. Hanani, N. Sauer, & J. Schönheim, eds.), Gordon and Breach, New York; pp. 311–313. (Cited in §7.2.)

**S. Olariu, J. L. Schwing, & J. Zhang (1995).**
Interval graph problems on reconfigurable meshes. *ORSA J. Comput.* **7**, 333–348. (Cited in §3.4.3.)

**R. J. Opsut (1982).**
On the computation of the competition number of a graph. *SIAM J. Alg. Discrete Methods* **3**, 420–428. (Cited in §4.2.)

**R. J. Opsut & F. S. Roberts (1981).**
On the fleet maintenance, mobile radio frequency, task assignment,

and traffic phasing problems. In *The Theory and Applications of Graphs* (G. Chartrand, Y. Alavi, D. L. Goldsmith, L. Lesniak-Foster, & D. R. Lick, eds.), Wiley, New York; pp. 479–492. (Cited in §3.1.)

**E. T. Ordman (1989).**
Minimal threshold separators and memory requirements for synchronization. *SIAM J. Comput.* **18**, 152–165. (Cited in §5.4.)

**B. S. Panda (1996).**
New linear time algorithms for generating perfect elimination orderings of chordal graphs. *Inform. Process. Lett.* **58**, 111–115. (Cited in §2.2.)

**B. S. Panda & S. P. Mohanty (1995).**
Intersection graphs of vertex disjoint paths in a tree. *Discrete Math.* **146**, 179–209. (Cited in §7.1.)

**A. Parra & P. Scheffler (1995).**
How to use the minimal separators of a graph for its chordal triangulation. In *Automata, Languages and Programming* (Z. Gülöp & F. Gécseg, eds.) [*Lecture Notes in Computer Science* **944**], Springer, Berlin; pp. 123–134. (Cited in §2.4.2.)

**A. Parra & P. Scheffler (1997).**
Characterizations and algorithmic applications of chordal graph embeddings. *Discrete Appl. Math.* **79**, 171–188. (Cited in §2.4.2.)

**J. Pearl (1988).**
*Probabilistic Reasoning in Intelligent Systems: Networks of Plausible Inference.* Morgan Kaufmann, San Mateo, CA. (Cited in §2.4.4.)

**I. Pe'er & R. Shamir (1995).**
Interval graphs with side (and size) constraints. In *Algorithms — ESA '95* (P. Spirakis, ed.) [*Lecture Notes in Computer Science* **979**], Springer, Berlin; pp. 142–154. (Cited in §3.3.)

**B. W. Peyton, A. Pothen, & X. Yuan (1995).**
A clique tree algorithm for partitioning a chordal graph into transitive subgraphs. *Linear Algebra Appl.* **223/224**, 553–588. (Cited in §7.9.)

**S. L. Pimm (1991).**
*The Balance of Nature.* Univ. of Chicago Press. (Cited in §4.3.)

**A. Pnueli, A. Lempel, & S. Even (1971).**
Transitive orientation of graphs and identification of permutation graphs. *Canad. J. Math.* **23**, 160–175. (Cited in §7.4.)

**B. Pondělíček (to appear).**
Chordal intersection graphs of bands. *Czechoslovak Math. J.* (Cited in §7.8.)

**E. Prisner (1989).**
A characterization of interval catch graphs. *Discrete Math.* **73**, 285–289. (Cited in §7.2.)

**E. Prisner (1992).**
Representing triangulated graphs in stars. *Abh. Math. Sem. Univ. Hamburg* **62**, 29–41. (Cited in §2.1.)

**E. Prisner (1993).**
Hereditary clique-Helly graphs. *J. Combin. Math. Combin. Comput.* **14**, 216–220. (Cited in §7.5.)

**E. Prisner (1994).**
Intersection-representation by connected subgraphs of some $n$-cyclomatic graph. *Ars Combin.* **37**, 241–256. (Cited in §7.1.)

**E. Prisner (1995).**
*Graph Dynamics.* [*Pitman Research Notes in Mathematics* **338**], Longman, London. (Cited in the Preface, beginning of §4, §7.5, 7.10.)

**E. Prisner (1996a).**
Line graphs and generalizations—A survey. In *Surveys in Graph Theory* (G. Chartrand & M. Jacobson, eds.), [Special volume, **116**, *Congr. Numer.*], Utilitas Mathematica, Winnipeg; pp. 193–229. (Cited in §1.5.)

**E. Prisner (1996b).**
A note on powers and proper circular-arc graphs. *J. Combin. Math. Combin. Comput.* **22**, 125–128. (Cited in §7.10.)

**E. Prisner (to appear).**
Intersection multigraphs of uniform hypergraphs. *Graphs Combin.* (Cited in §6.2.)

**M. Quest & G. Wegner (1990).**
Characterization of the graphs with boxicity $\leq 2$. *Discrete Math.* **81**, 187–192. (Cited in §7.1.)

**A. Quilliot (1988).**
On the problem of how to represent a graph taking into account an additional structure. *J. Combin. Theory Ser. B* **44**, 1–21. (Cited in §1.2.)

**C. W. Rasmussen (1994).**
Conditional graph completions. *Congr. Numer.* **103**, 183–192. (Cited in §7.8.)

**A. Raychaudhuri (1987).**
On powers of interval and unit interval graphs. *Congr. Numer.* **59**, 235–242. (Cited in §7.10.)

**A. Raychaudhuri (1988).**
Intersection number and edge clique graphs of chordal and strongly chordal graphs. *Congr. Numer.* **67**, 197–204. (Cited in §2.1, 7.12.)

**A. Raychaudhuri (1992a).**
On powers of strongly chordal and circular arc graphs. *Ars Combin.* **34**, 147–160. (Cited in §7.10.)

**A. Raychaudhuri (1992b).**
Optimal multiple interval assignments in frequency assignment and traffic phasing. *Discrete Appl. Math.* **40**, 319–332. (Cited in §7.1.)

**A. Raychaudhuri & F. S. Roberts (1985).**
Generalized competition graphs and their applications. *Methods Oper. Res.* **49**, pp. 295–311. (Cited in §4.2.)

**L. Rédei (1934).**
Ein kombinatorischer Satz. *Acta Litt. Szeged* **7**, 39–43. (Cited in §3.1.)

**P. L. Renz (1970).**
Intersection representation of graphs by arcs. *Pacific J. Math.* **34**, 501–510. (Cited in §7.1.)

**J. Riguet (1951).**
Les relations de Ferrers. *C. R. Acad. Sci. Paris* **232**, 1729–1730. (Cited in §5.3.)

**C. S. Rim & K. Nakajima (1995).**
On rectangle intersection and overlap graphs. *IEEE Trans. Circuits Systems* I *Fund. Theory Appl.* **42**, 549–553. (Cited in §7.1.)

**F. S. Roberts (1968).**
*Representations of Indifference Relations.* Ph.D. thesis, Stanford University, Stanford, CA. (Cited in §3.3, 7.1.)

**F. S. Roberts (1969a).**
Indifference graphs. In *Proof Techniques in Graph Theory* (F. Harary, ed.), Academic Press, New York; pp. 139–146. (Cited in §3.3, 3.4.2, 6.3, 7.2.)

**F. S. Roberts (1969b).**
On the boxicity and cubicity of a graph. In *Recent Progress in Combinatorics* (W. T. Tutte, ed.), Academic Press, New York; pp. 301–310. (Cited in §7.1.)

**F. S. Roberts (1971).**
On the compatibility between a graph and a simple order. *J. Combin. Theory* **11**, 28–38. (Cited in §3.4.2, 7.2.)

**F. S. Roberts (1976).**
*Discrete Mathematical Models, with Applications to Social, Biological and Environmental Problems.* Prentice–Hall, Englewood Cliffs, NJ. (Cited in Preface, beginning of §3, §3.1, 3.4.2, 4.2.)

**F. S. Roberts (1978a).**
Food webs, competition graphs, and the boxicity of ecological phase space. In *Theory and Applications of Graphs* (Y. Alavi & D. Lick, eds.) [*Lecture Notes in Mathematics* **642**], Springer, New York; pp. 477–490. (Cited in Preface, §3.1, 3.4.2, 4.2.)

**F. S. Roberts (1978b).**
*Graph Theory and Its Applications to Problems of Society* [*CBMS-NSF Regional Conference Series in Applied Mathematics* **29**]. Society for Industrial and Applied Mathematics, Philadelphia. (Cited in Preface, §3.4.2, 4.2.)

**F. S. Roberts (1979).**
*Measurement Theory with Applications to Decisionmaking, Utility and the Social Sciences.* [*Encyclopedia of Mathematics and its Applications* **7**.] Addison–Wesley, Reading, MA. (Cited in Preface, §3.4.2.)

**F. S. Roberts (1985).**
Applications of edge coverings by cliques. *Discrete Appl. Math.* **10**, 93–109. (Cited in beginning of §1.)

**F. S. Roberts (1989).**
Applications of combinatorics and graph theory to the biological and social sciences: Seven fundamental ideas. In *Applications of Combinatorics and Graph Theory to the Biological and Social Sciences.* (F. S. Roberts, ed.) [*IMA Volumes in Mathematics and Its Applications* **17**.], Springer, New York; pp. 1–37. (Cited in §7.1.)

**F. S. Roberts & J. H. Spencer (1971).**
A characterization of clique graphs. *J. Combin. Theory Ser. B* **10**, 102–108. (Cited in §1.4, 1.6, 6.2.)

**F. S. Roberts & J. E. Steif (1983).**
A characterization of competition graphs of arbitrary digraphs. *Discrete Appl. Math.* **6**, 323–326. (Cited in §4.2, 6.1.)

**N. Robertson & P. D. Seymour (1985).**
Graph minors—A survey. In *Surveys in Combinatorics.* (I. Anderson, ed.) Cambridge University Press; pp. 153–171. (Cited in §7.1.)

**D. J. Rose (1970).**
Triangulated graphs and the elimination process. *J. Math. Anal. Appl.* **32**, 597–609. (Cited in §2.2, 2.4.3, 7.2.)

**D. J. Rose (1972).**
A graph-theoretic study of the numerical solution of sparse positive definite systems of linear equations. In *Graph Theory and Computing* (R. C. Read, ed.), Academic Press, New York; pp. 183–217. (Cited in §2.4.3.)

**D. J. Rose & R. E. Tarjan (1975).**
Algorithmic aspects of vertex elimination. In *Seventh Annual ACM Symposium on Theory of Computing*, Assoc. Comput. Mach., New York; pp. 245–254. (Cited in §2.4.1.)

**D. J. Rose, R. E. Tarjan, & G. S. Lueker (1976).**
Algorithmic aspects of vertex elimination on graphs. *SIAM J. Comput.* **5**, 266–283. (Cited in §2.4.1.)

**N. D. Roussopoulos (1973).**
A max$\{m, n\}$ algorithm for determining the graph $H$ from its line graph $G$. *Inform. Process. Lett.* **2**, 108–112. (Cited in §1.5.)

**H. J. Ryser (1969).**
Combinatorial configurations. *SIAM J. Appl. Math.* **17**, 593–602. (Cited in §2.3.)

**H. Sachs (1994).**
Coin graphs, polyhedra, and conformal mapping. *Discrete Math.* **134**, 133–138. (Cited in §7.1.)

**T. J. Santner & D. E. Duffy (1989).**
*The Statistical Analysis of Discrete Data.* Springer, New York. (Cited in §2.4.4.)

**B. K. Sanyal & M. K. Sen (1996).**
New characterizations of graphs represented by intervals. *J. Graph Theory* **22**, 297–303. (Cited in §7.2.)

**E. R. Scheinerman (1984).**
*Intersection classes and multiple intersection parameters of a graph.* Ph.D. thesis, Princeton University, Princeton, NJ. (Cited in §7.1.)

**E. R. Scheinerman (1985a).**
Characterizing intersection classes of graphs. *Discrete Math.* **55**, 185–193. (Cited in §1.2.)

**E. R. Scheinerman (1985b).**
Irrepresentability by multiple intersection, or why the interval number is unbounded. *Discrete Math.* **55**, 195–211. (Cited in §7.1.)

**E. R. Scheinerman (1985c).**
Characterization and computational complexity questions for representation classes of graphs. *Congr. Numer.* **49**, 195–204. (Cited in §1.2.)

**E. R. Scheinerman (1986).**
On the structure of hereditary classes of graphs. *J. Graph Theory* **10**, 545–551. (Cited in §1.2.)

**E. R. Scheinerman (1988a).**
On the interval number of a chordal graph. *J. Graph Theory* **12**, 311–316. (Cited in §7.1.)

**E. R. Scheinerman (1988b).**
Random interval graphs. *Combinatorica* **8**, 357–371. (Cited in §7.8.)

**E. R. Scheinerman (1990a).**
On the interval number of random graphs. *Discrete Math.* **82**, 105–109. (Cited in §7.8.)

**E. R. Scheinerman (1990b).**
An evolution of interval graphs. *Discrete Math.* **82**, 287–302. (Cited in §7.8.)

**E. R. Scheinerman & D. B. West (1983).**
The interval number of a planar graph: Three intervals suffice. *J. Combin. Theory Ser. B* **35**, 224–239. (Cited in §7.1.)

**D. D. Scott (1986).**
Posets with interval upper bound graphs. *Order* **3**, 269–281. (Cited in §4.4.)

**M. Sen, S. Das, A. B. Roy, & D. B. West (1989).**
Interval digraphs: An analogue of interval graphs. *J. Graph Theory* **13**, 189–202. (Cited in §7.2.)

**M. Sen, S. Das, & D. B. West (1989).**
Circular-arc digraphs: A characterization. *J. Graph Theory* **13**, 581–592. (Cited in §7.2.)

**M. Sen, S. Das, & D. B. West (1992).**
Representing digraphs by arcs of a circle. *Sankhya, Ser. A— "special issue"* **54**, 421–427. (Cited in §7.2.)

**M. Sen & B. K. Sanyal (1994).**
Indifference digraphs: A generalization of indifference graphs and semi-orders. *SIAM J. Discrete Math.* **7**, 157–165. (Cited in §7.2.)

**M. Sen, B. K. Sanyal, & D. B. West (1995).**
Representing digraphs using intervals or circular arcs. *Discrete Math.* **147**, 235–245. (Cited in §7.2, 7.6.)

**L. Sheng, C. Wang, & P. Zhang (to appear).**
Tagged probe interval graphs. (Cited in §3.4.1.)

**L. N. Shevrin & A. J. Ovsyannikov (1983).**
Semigroups and their subsemigroup lattices. *Semigroup Forum* **27**, 1–154. (Cited in §7.8.)

**Y. Shibata (1988).**

On the tree representation of chordal graphs. *J. Graph Theory* **12**, 421–428. (Cited in §2.1.)

**D. Shier (1984).**

Some aspects of perfect elimination orderings in chordal graphs. *Discrete Appl. Math.* **7**, 325–331. (Cited in §2.2.)

**R. Shull & A. N. Trenk (1997).**

Unit and proper bitolerance digraphs. *J. Graph Theory* **24**, 193–199. (Cited in §6.3.)

**K. Simon (1991).**

A new simple linear algorithm to recognize interval graphs. In *Computational Geometry—Methods, Algorithms and Applications* (H. Bieri & H. Noltemeier, eds.), [*Lecture Notes in Computer Science* **553**], Springer, Berlin; pp. 289–308. (Cited in §3.1.)

**K. Simon (1995).**

A note on lexicographic breadth first search for chordal graphs. *Inform. Process. Lett.* **54**, 249–251. (Cited in §2.2.)

**D. J. Skrien (1982).**

A relationship between triangulated graphs, comparability graphs, proper interval graphs, proper circular-arc graphs, and nested interval graphs. *J. Graph Theory* **6**, 309–316. (Cited in §7.1, 7.9.)

**D. J. Skrien (1984).**

Chronological orderings of interval graphs. *Discrete Appl. Math.* **8**, 69–83. (Cited in §3.3, 5.2.)

**P. J. Slater (1976).**

A note on pseudointersection graphs. *J. Res. Nat. Bur. Standards Section B* **80**, 441–445. (Cited in §1.3.)

**P. J. Slater (1978).**

A characterization of SOFT hypergraphs. *Canad. Math. Bull.* **21**, 335–337. (Cited in §2.3.)

**M. Ślusarek (1989).**

A coloring algorithm for interval graphs. In *Mathematical Foundations of Computer Science* 1989 (A. Kreczmar & G. Mirkowska, eds.), [*Lecture Notes in Computer Science* **379**], Springer, Berlin, pp. 471–480. (Cited in §3.4.)

**M. Ślusarek (1995).**
 Optimal on-line coloring of circular arc graphs. *RAIRO Inform. Théor. Appl.* **29**, 423–429. (Cited in §3.4.)

**T. P. Speed & H. T. Kiiveri (1986).**
 Gaussian Markov distributions over finite fields. *Ann. Statist.* **14**, 138–150. (Cited in §2.4.4.)

**J. Spinrad (1993).**
 Doubly lexical orderings of dense 0-1 matrices. *Inform. Process. Lett.* **45**, 229–235. (Cited in §7.3.)

**J. Spinrad (1994).**
 Recognition of circle graphs. *J. Algorithms* **16**, 264–282. (Cited in §7.4.)

**J. Spinrad (1995).**
 Nonredundant 1's in Γ-free matrices. *SIAM J. Discrete Math.* **8**, 251–257. (Cited in §7.3.)

**J. Spinrad & R. Sritharan (1995).**
 Algorithms for weakly triangulated graphs. *Discrete Appl. Math.* **59**, 181–191. (Cited in §7.3, 7.8.)

**J. Spinrad, G. Vijayan, & D. B. West (1987).**
 An improved edge bound on the interval number of a graph. *J. Graph Theory* **11**, 447–449. (Cited in §7.1.)

**G. Steiner (1996).**
 The recognition of indifference digraphs and generalized semiorders. *J. Graph Theory* **21**, 235–241. (Cited in §7.2.)

**S. K. Stueckle, B. L. Piazza, & R. D. Ringeisen (1995).**
 A circular-arc characterization of certain rectilinear drawings. *J. Graph Theory* **20**, 71–76. (Cited in §7.1.)

**G. Sugihara (1984).**
 Graph theory, homology, and food webs. In *Population Biology* (S. A. Levin, ed.) [*Proceedings of Symposia in Applied Mathematics* **30**], American Mathematical Society, Providence, RI; pp. 83–101. (Cited in §4.3.)

**D. P. Sumner (1973).**
Point determination in graphs. *Discrete Math.* **5**, 179–187. (Cited in §1.3.)

**M. M. Sysło (1985).**
Triangulated edge intersection graphs of paths in a tree. *Discrete Math.* **55**, 217–220. (Cited in §7.1.)

**J. L. Szwarcfiter (1997).**
Recognizing clique-Helly graphs. *Ars Combin.* **45**, 29–32. (Cited in §7.5.)

**J. L. Szwarcfiter & C. F. Bornstein (1994).**
Clique graphs of chordal and path graphs. *SIAM J. Discrete Math.* **7**, 331–336. (Cited in §7.5.)

**K. Tanaka (1983).**
Tree-structured data organization with consecutive retrieval property. In *Data Base File Organization, Theory and Applications of the Consecutive Retrieval Property* (S. P. Ghosh, Y. Kambayashi, & W. Lipski, eds.), Academic Press, New York; pp. 271–276. (Cited in §3.4.3.)

**R. E. Tarjan (1983).**
*Data Structures and Network Algorithms.* Society for Industrial and Applied Mathematics, Philadelphia. (Cited in §2.2.)

**R. E. Tarjan & M. Yannakakis (1984).**
Simple linear-time algorithms to test chordality of graphs, test acyclicity of hypergraphs, and selectively reduce acyclic hypergraphs. *SIAM J. Comput.* **13**, 566–579. (Cited in §2.2.)

**C. Thomassen (1986).**
Interval representations of planar graphs. *J. Combin. Theory, Series B* **40**, 9–20. (Cited in §7.1.)

**G. T. Toussaint (1988).**
A graph-theoretic primal sketch. In *Computer Morphology* (G. T. Toussain, ed.), Elsevier, Amsterdam; pp. 229–260. (Cited in §7.11.)

**W. T. Trotter (1992).**
*Combinatorics and partially ordered sets. Dimension theory.* Johns Hopkins University Press, Baltimore, MD. (Cited in Preface.)

**W. T. Trotter & F. Harary (1978).**
On double and multiple interval graphs. *J. Graph Theory* **2**, 137–142. (Cited in §7.1.)

**W. T. Trotter & J. I. Moore (1976).**
Characterization problems for graphs, partially ordered sets, lattices, and families of sets. *Discrete Math.* **16**, 361–381. (Cited in §3.2.)

**W. T. Trotter & D. B. West (1987).**
Point boxicity of graphs. *Discrete Math.* **64**, 105–107. (Cited in §7.1.)

**D. S. Troxell (1995).**
On properties of unit interval graphs with a perceptual motivation. *Math. Social Sci.* **30**, 1–22. (Cited in §3.4.2.)

**M. Tsuchiya (1994).**
On antichain intersection numbers, total clique covers and regular graphs. *Discrete Math.* **127**, 305–318. (Cited in §1.3.)

**A. C. Tucker (1971).**
Matrix characterizations of circular-arc graphs. *Pacific J. Math.* **39**, 535–545. (Cited in §7.1.)

**A. C. Tucker (1972).**
A structure theorem for the consecutive 1's property. *J. Combin. Theory, Ser. B* **12**, 153–162. (Cited in §3.2.)

**A. C. Tucker (1974).**
Structure theorems for some circular-arc graphs. *Discrete Math.* **7**, 167–195. (Cited in §7.1.)

**A. C. Tucker (1978).**
Circular arc graphs: New uses and a new algorithm. In *Theory and Applications of Graphs* (Y. Alavi & D. Lick, eds.) [*Lecture Notes in Mathematics* **642**], Springer, New York; pp. 580–589. (Cited in §7.1.)

**R. I. Tyškevič & A. A. Černjak (1978a).**
Unigraphs. I. *Vestsĭ Akad. Navuk BSSR Ser. Fĭz.-Mat. Navuk* no. 5, 5–11, 141. (Cited in §2.5.)

**R. I. Tyškevič & A. A. Černjak (1978b).**
Unigraphs. II. *Vestsĭ Akad. Navuk BSSR Ser. Fĭz.-Mat. Navuk* no. 1, 5–12, 138. (Cited in §2.5.)

**R. I. Tyškevič & A. A. Černjak (1979).**
Unigraphs. III. *Vestsī Akad. Navuk BSSR Ser. Fīz.-Mat. Navuk* no. 2, 5–11, 138. (Cited in §2.5.)

**W. D. Wallis & J. Wu (1995).**
Squares, clique graphs, and chordality. *J. Graph Theory* **20**, 37–45. (Cited in §7.10.)

**W. D. Wallis & G.-H. Zhang (1990).**
On maximal clique irreducible graphs. *J. Combin. Math. Combin. Comput.* **8**, 187–193. (Cited in §7.5.)

**J. R. Walter (1978).**
Representations of chordal graphs as subtrees of a tree. *J. Graph Theory* **2**, 265–267. (Cited in §2.1.)

**H. Wan, E. Lee, C. Wang, & P. Zhang (to appear).**
$k$-partite probe interval graphs: A computational model for physical mapping of DNA. (Cited in §3.4.1.)

**C. Wang (1992).**
On critical graphs for Opsut's conjecture. *Ars Combin.* **34**, 183–203. (Cited in §4.2.)

**C. Wang (1994).**
A subgraph problem from restriction maps of DNA. *J. Comput. Biology* **3**, 227–234. (Cited in §3.4.1.)

**C. Wang (1995a).**
Competition graphs and resource graphs of digraphs. *Ars Combin.* **40**, 3–48. (Cited in §4.2.)

**C. Wang (1995b).**
Competitive inheritance and limitedness of graphs. *J. Graph Theory* **19**, 353–366. (Cited in §4.2.)

**M. S. Waterman (1995).**
*Introduction to Computational Biology.* Chapman and Hall, London. (Cited in §3.4.1.)

**G. Wegner (1967).**
*Eigenschaften der Nerven Homologisch-einfacher Familien in $R^n$.* Ph.D. thesis, Göttingen. (Cited in §3.3, 7.1.)

**D. B. West (1996).**
Introduction to Graph Theory. Prentice–Hall, Upper Saddle River, NJ. (Cited in Preface, §1.1.)

**D. B. West (1998).**
Short proofs for interval digraphs. Discrete Math. **178**, 287–292. (Cited in §7.2.)

**D. B. West & D. B. Shmoys (1984).**
Recognizing graphs with fixed interval number is NP-complete. Discrete Appl. Math. **8**, 295–305. (Cited in §7.1.)

**H. Whitney (1932).**
Congruent graphs and the connectivity of graphs. Amer. J. Math. **54**, 150–168. (Cited in §1.5.)

**J. Whittaker (1990).**
Graphical Models in Applied Multivariate Statistics. Wiley, New York. (Cited in §2.4.4.)

**T. D. Wickens (1989).**
Multiway Contingency Tables Analysis for the Social Sciences. Erlbaum, Hillsdale, NJ. (Cited in §2.4.4.)

**E. S. Wolk (1962).**
The comparability graph of a tree. Proc. Amer. Math. Soc. **13**, 789–795. (Cited in §7.7, 7.9.)

**E. S. Wolk (1965).**
A note on "The comparability graph of a tree." Proc. Amer. Math. Soc. **16**, 17–20. (Cited in §7.9.)

**J.-H. Yan, J.-J. Chen & G. J. Chang (1996).**
Quasi-threshold graphs. Discrete Appl. Math. **69**, 247–255. (Cited in §7.9.)

**T. Zamfirescu (1973/74).**
A characterization of Hamiltonian graphs. Atti Acad. Sci. Istit. Bologna Cl. Sci. Fis. Rend. **13**, 39–40. (Cited in §1.2.)

**B. Zelinka (1975a).**
Intersection graphs of finite abelian groups. Czechoslovak Math. J. **25** (100), 171–174. (Cited in §7.8.)

**B. Zelinka (1975b).**
Intersection graphs of graphs. *Mat. Časopis Sloven. Akad. Vied* **25**, 129–133. (Cited in §7.8.)

**P. Zhang (to appear).**
Probe interval graphs and their application to physical mapping of DNA. (Cited in §3.4.1.)

**A. A. Zykov (1987).**
*Fundamentals of Graph Theory.* (English translation, 1990.) BCS Associates, Moscow, ID. (Cited in Preface.)

# Index